ISBN 978-1-332-43221-9
PIBN 10426109

"SQUARING THE CIRCLE"

A HISTORY OF THE PROBLEM

CAMBRIDGE UNIVERSITY PRESS
London: FETTER LANE, E.C.
C. F. CLAY, Manager

Edinburgh: 100, PRINCES STREET
Berlin: A. ASHER AND CO.
Leipzig: F. A. BROCKHAUS
New York: G. P. PUTNAM'S SONS
Bombay and Calcutta: MACMILLAN AND CO., Ltd.
Toronto: J. M. DENT AND SONS, Ltd.
Tokyo: THE MARUZEN-KABUSHIKI-KAISHA

"SQUARING THE CIRCLE"

A HISTORY OF THE PROBLEM

BY

E. W. HOBSON, Sc.D., LL.D., F.R.S.

SADLEIRIAN PROFESSOR OF PURE MATHEMATICS, AND FELLOW OF CHRIST'S COLLEGE, IN THE UNIVERSITY OF CAMBRIDGE

Cambridge:
at the University Press
1913

Cambridge:
PRINTED BY JOHN CLAY, M.A.
AT THE UNIVERSITY PRESS

PREFACE

IN the Easter Term of the present year I delivered a short course of six Professorial Lectures on the history of the problem of the quadrature of the circle, in the hope that a short account of the fortunes of this celebrated problem might not only prove interesting in itself, but might also act as a stimulant of interest in the more general history of Mathematics. It has occurred to me that, by the publication of the Lectures, they might perhaps be of use, in the same way, to a larger circle of students of Mathematics.

The account of the problem here given is not the result of any independent historical research, but the facts have been taken from the writings of those authors who have investigated various parts of the history of the problem.

The works to which I am most indebted are the very interesting book by Prof. F. Rudio entitled "Archimedes, Huygens, Lambert, Legendre. Vier Abhandlungen über die Kreismessung" (Leipzig, 1892), and Sir T. L. Heath's treatise "The works of Archimedes" (Cambridge, 1897). I have also made use of Cantor's "Geschichte der Mathematik," of Vahlen's "Konstruktionen und Approximationen" (Leipzig, 1911), of Yoshio Mikami's treatise "The development of Mathematics in China and Japan" (Leipzig, 1913), of the translation by T. J. McCormack (Chicago, 1898) of H. Schubert's "Mathematical Essays and Recreations," and of the article "The history and transcendence of π" written by Prof. D. E. Smith which appeared in the "Monographs on Modern Mathematics" edited by Prof. J. W. A. Young. On special points I have consulted various other writings.

E. W. H.

CHRIST'S COLLEGE, CAMBRIDGE.
October, 1913.

CONTENTS

CHAPTER I

CHAPTER II

CHAPTER III

CHAPTER IV

CHAPTER I

GENERAL ACCOUNT OF THE PROBLEM

A GENERAL survey of the history of thought reveals to us the fact of the existence of various questions that have occupied the almost continuous attention of the thinking part of mankind for long series of centuries. Certain fundamental questions presented themselves to the human mind at the dawn of the history of speculative thought, and have maintained their substantial identity throughout the centuries, although the precise terms in which such questions have been stated have varied from age to age in accordance with the ever varying attitude of mankind towards fundamentals. In general, it may be maintained that, to such questions, even after thousands of years of discussion, no answers have been given that have permanently satisfied the thinking world, or that have been generally accepted as final solutions of the matters concerned. It has been said that those problems that have the longest history are the insoluble ones.

If the contemplation of this kind of relative failure of the efforts of the human mind is calculated to produce a certain sense of depression, it may be a relief to turn to certain problems, albeit in a more restricted domain, that have occupied the minds of men for thousands of years, but which have at last, in the course of the nineteenth century, received solutions that we have reasons of overwhelming cogency to regard as final. Success, even in a comparatively limited field, is some compensation for failure in a wider field of endeavour. Our legitimate satisfaction at such exceptional success is but slightly qualified by the fact that the answers ultimately reached are in a certain sense of a negative character. We may rest contented with proofs that these problems, in their original somewhat narrow form, are insoluble, provided we attain, as is actually the case in some celebrated instances, to a complete comprehension of the grounds, resting upon a thoroughly established theoretical basis, upon which our final conviction of the insolubility of the problems is founded.

The three celebrated problems of the quadrature of the circle, the trisection of an angle, and the duplication of the cube, although all of them are somewhat special in character, have one great advantage for the purposes of historical study, *viz.* that their complete history as scientific problems lies, in a completed form, before us. Taking the first of these problems, which will be here our special subject of study, we possess indications of its origin in remote antiquity, we are able to follow the lines on which the treatment of the problem proceeded and changed from age to age in accordance with the progressive development of general Mathematical Science, on which it exercised a noticeable reaction. We are also able to see how the progress of endeavours towards a solution was affected by the intervention of some of the greatest Mathematical thinkers that the world has seen, such men as Archimedes, Huyghens, Euler, and Hermite. Lastly, we know when and how the resources of modern Mathematical Science became sufficiently powerful to make possible that resolution of the problem which, although negative, in that the impossibility of the problem subject to the implied restrictions was proved, is far from being a mere negation, in that the true grounds of the impossibility have been set forth with a finality and completeness which is somewhat rare in the history of Science.

If the question be raised, why such an apparently special problem, as that of the quadrature of the circle, is deserving of the sustained interest which has attached to it, and which it still possesses, the answer is only to be found in a scrutiny of the history of the problem, and especially in the closeness of the connection of that history with the general history of Mathematical Science. It would be difficult to select another special problem, an account of the history of which would afford so good an opportunity of obtaining a glimpse of so many of the main phases of the development of general Mathematics; and it is for that reason, even more than on account of the intrinsic interest of the problem, that I have selected it as appropriate for treatment in a short course of lectures.

Apart from, and alongside of, the scientific history of the problem, it has a history of another kind, due to the fact that, at all times, and almost as much at the present time as formerly, it has attracted the attention of a class of persons who have, usually with a very inadequate equipment of knowledge of the true nature of the problem or of its history, devoted their attention to it, often with passionate enthusiasm. Such persons have very frequently maintained, in the face of all efforts at refutation made by genuine Mathematicians, that they had obtained a solution of the problem. The solutions propounded by the circle squarer exhibit every grade of skill, varying from the most futile attempts, in which the writers shew an utter lack of power to reason correctly, up to approximate solutions the construction of which required much ingenuity on the part of their inventor. In some cases it requires an effort of sustained attention to find out the precise point in the demonstration at which the error occurs, or in which an approximate determination is made to do duty for a theoretically exact one. The psychology of the scientific crank is a subject with which the officials of every Scientific Society have some practical acquaintance. Every Scientific Society still receives from time to time communications from the circle squarer and the trisector of angles, who often make amusing attempts to disguise the real character of their essays. The solutions propounded by such persons usually involve some misunderstanding as to the nature of the conditions under which the problems are to be solved, and ignore the difference between an approximate construction and the solution of the ideal problem. It is a common occurrence that such a person sends his solution to the authorities of a foreign University or Scientific Society, accompanied by a statement that the men of Science of the writer's own country have entered into a conspiracy to suppress his work, owing to jealousy, and that he hopes to receive fairer treatment abroad. The statement is not infrequently accompanied with directions as to the forwarding of any prize of which the writer may be found worthy by the University or Scientific Society addressed, and usually indicates no lack of confidence that the bestowal of such a prize has been amply deserved as the fit reward for the final solution of a problem which has baffled the efforts of a great multitude of predecessors in all ages. A very interesting detailed account of the peculiarities of the circle squarer, and of the futility of attempts on the part of Mathematicians to convince him of his errors, will be found in Augustus De Morgan's *Budget of Paradoxes*. As early as the time of the Greek Mathematicians circle-squaring occupied the attention of non-Mathematicians; in fact the Greeks had a special word to denote this kind of activity, *viz.* τετραγωνίζειν, which means to occupy oneself with the quadrature. It is interesting to remark that, in the year 1775, the Paris Academy found it necessary to protect its officials against the waste of time and energy involved in examining the efforts of circle squarers. It passed a resolution, which appears in the Minutes of the Academy*, that no more solutions were to be examined of the problems of the duplication of the cube, the trisection of the angle, the quadrature of the circle, and that the same resolution should apply to machines for exhibiting perpetual motion. An account of the reasons which led to the adoption of this resolution, drawn up by Condorcet, who was then the perpetual Secretary of the Academy, is appended. It is interesting to remark the strength of the conviction of Mathematicians that the solution of the problem is impossible, more than a century before an irrefutable proof of the correctness of that conviction was discovered.

The popularity of the problem among non-Mathematicians may seem to require some explanation. No doubt, the fact of its comparative obviousness explains in part at least its popularity; unlike many Mathematical problems, its nature can in some sense be understood by anyone; although, as we shall presently see, the very terms in which it is usually stated tend to suggest an imperfect apprehension of its precise import. The accumulated celebrity which the problem attained, as one of proverbial difficulty, makes it an irresistible attraction to men with a certain kind of mentality. An exaggerated notion of the gain which would accrue to mankind by a solution of the problem has at various times been a factor in stimulating the efforts of men with more zeal than knowledge. The man of mystical tendencies has been attracted to the problem by a vague idea that its solution would, in some dimly discerned manner, prove a key to a knowledge of the inner connections of things far beyond those with which the problem is immediately connected.

Statement of the problem

The fact was well known to the Greek Geometers that the problems of the quadrature and the rectification of the circle are equivalent problems. It was in fact at an early time established that the ratio of the length of a complete circle to the diameter has a definite value equal to that of the area of the circle to that of a square of which the radius is side. Since the time of Euler this ratio has always been denoted by the familiar notation π. The problem of "squaring the circle" is roughly that of constructing a square of which the area is equal to that enclosed by the circle. This is then equivalent to the problem of the rectification of the circle, *i.e.* of the determination of a

* *Histoire de l'Académie royale*, année 1775, p. 61.

straight line, of which the length is equal to that of the circumference of the circle. But a problem of this kind becomes definite only when it is specified what means are to be at our disposal for the purpose of making the required construction or determination; accordingly, in order to present the statement of our problem in a precise form, it is necessary to give some preliminary explanations as to the nature of the postulations which underlie all geometrical procedure.

The Science of Geometry has two sides; on the one side, that of practical or physical Geometry, it is a physical Science concerned with the actual spatial relations of the extended bodies which we perceive in the physical world. It was in connection with our interests, of a practical character, in the physical world, that Geometry took its origin. Herodotus ascribes its origin in Egypt to the necessity of measuring the areas of estates of which the boundaries had been obliterated by the inundations of the Nile, the inhabitants being compelled, in order to settle disputes, to compare the areas of fields of different shapes. On this side of Geometry, the objects spoken of, such as points, lines, &c., are physical objects; a point is a very small object of scarcely perceptible and practically negligible dimensions; a line is an object of small, and for some purposes negligible, thickness; and so on. The constructions of figures consisting of points, straight lines, circles, &c., which we draw, are constructions of actual physical objects. In this domain, the possibility of making a particular construction is dependent upon the instruments which we have at our disposal.

On the other side of the subject, Geometry is an abstract or rational Science which deals with the relations of objects that are no longer physical objects, although these ideal objects, points, straight lines, circles, &c., are called by the same names by which we denote their physical counterparts. At the base of this rational Science there lies a set of definitions and postulations which specify the nature of the relations between the ideal objects with which the Science deals. These postulations and definitions were suggested by our actual spatial perceptions, but they contain an element of absolute exactness which is wanting in the rough data provided by our senses. The objects of abstract Geometry possess in absolute precision properties which are only approximately realized in the corresponding objects of physical Geometry. In every department of Science there exists in a greater or less degree this distinction between the abstract or rational side and the physical or concrete side; and the progress of each department of Science involves a continually increasing amount of rationalization. In Geometry the passage from a purely empirical treatment to the setting up of a rational Science proceeded by much more rapid stages than in other cases. We have in the Greek Geometry, known to us all through the presentation of it given in that oldest of all scientific text books, Euclid's *Elements of Geometry*, a treatment of the subject in which the process of rationalization has already reached an advanced stage. The possibility of solving a particular problem of determination, such as the one we are contemplating, as a problem of rational Geometry, depends upon the postulations that are made as to the allowable modes of determination of new geometrical elements by means of assigned ones. The restriction in practical Geometry to the use of specified instruments has its counterpart in theoretical Geometry in restrictions as to the mode in which new elements are to be determined by means of given ones. As regards the postulations of rational Geometry in this respect there is a certain arbitrariness corresponding to the more or less arbitrary restriction in practical Geometry to the use of specified instruments.

The ordinary obliteration of the distinction between abstract and physical Geometry is furthered by the fact that we all of us, habitually and almost necessarily, consider both aspects of the subject at the same time. We may be thinking out a chain of reasoning in abstract Geometry, but if we draw a figure, as we usually must do in order to fix our ideas and prevent our attention from wandering owing to the difficulty of keeping a long chain of syllogisms in our minds, it is excusable if we are apt to forget that we are not in reality reasoning about the objects in the figure, but about objects which are their idealizations, and of which the objects in the figure are only an imperfect representation. Even if we only visualize, we see the images of more or less gross physical objects, in which various qualities irrelevant for our specific purpose are not entirely absent, and which are at best only approximate images of those objects about which we are reasoning.

It is usually stated that the problem of squaring the circle, or the equivalent one of rectifying it, is that of constructing a square of an area equal to that of the circle, or in the latter case of constructing a straight line of length equal to that of the circumference, by a method which involves the use only of the compass and of the ruler as a single straight-edge. This mode of statement, although it indicates roughly the true statement of the problem, is decidedly defective in that it entirely leaves out of account the fundamental distinction between the two aspects of Geometry to which allusion has been made above. The compass and the straight-edge are physical objects by the use of which other objects can be constructed, *viz.* circles of small thickness, and lines which are approximately straight and very thin, made of ink or other material. Such instruments can clearly have no direct relation to theoretical Geometry, in which circles and straight lines are ideal objects possessing in absolute precision properties that are only approximately realized in the circles and straight lines that can be constructed by compasses and rulers. In theoretical Geometry, a restriction to the use of rulers and compasses, or of other instruments, must be replaced by corresponding postulations as to the allowable modes of determination of geometrical objects. We will see what these postulations really are in the case of Euclidean Geometry. Every Euclidean problem of construction, or as it would be preferable to say, every problem of determination, really consists in the determination of one or more points which shall satisfy prescribed conditions. We have here to consider the fundamental modes in which, when a number of points are regarded as given, or already determined, a new point is allowed to be determined.

Two of the fundamental postulations of Euclidean Geometry are that, having given two points A and B, then (1) a unique straight line (A, B) (the whole straight line, and not merely the segment between A and B) is determined such that A and B are incident on it, and (2) that a unique circle $A(B)$, of which A is centre and on which B is incident, is determined. The determinancy or assumption of existence of such straight lines and circles is in theoretical Geometry sufficient for the purposes of the subject. When we know that these objects, having known properties, exist, we may reason about them and employ them for the purposes of our further procedure; and that is sufficient for our purpose. The notion of drawing or constructing them by means of a straight-edge or compass has no relevance to abstract Geometry, but is borrowed from the language of practical Geometry.

A new point is determined in Euclidean Geometry exclusively in one of the three following ways:

Having given four points A, B, C, D, not all incident on the same straight line, then

(1) Whenever a point P exists which is incident both on (A, B) and on (C, D), that point is regarded as determinate.

(2) Whenever a point P exists which is incident both on the straight line (A, B) and on the circle $C(D)$, that point is regarded as determinate.

(3) Whenever a point P exists which is incident on both the circles $A(B)$, $C(D)$, that point is regarded as determinate.

The cardinal points of any figure determined by a Euclidean construction are always found by means of a finite number of successive applications of some or all of these rules (1), (2) and (3). Whenever one of these rules is applied it must be shewn that it does not fail to determine the point. Euclid's own treatment is sometimes defective as regards this requisite; as for example in the first proposition of his first book, in which it is not shewn that the circles intersect one another.

In order to make the practical constructions which correspond to these three Euclidean modes of determination, corresponding to (1) the ruler is required, corresponding to (2) both the ruler and the compass, and corresponding to (3) the compass only.

As Euclidean plane Geometry is concerned with the relations of points, straight lines, and circles only, it is clear that the above system of postulations, although arbitrary in appearance, is the system that the exigencies of the subject would naturally suggest. It may, however, be remarked that it is possible to develop Euclidean Geometry with a more restricted set of postulations. For example it can be shewn that all Euclidean constructions can be carried out by means of (3) alone*, without employing (1) or (2).

Having made these preliminary explanations we are now in a position to state in a precise form the ideal problem of "squaring the circle," or the equivalent one of the rectification of the circle.

The historical problem of "squaring the circle" is that of determining a square of which the area shall equal that of a given circle, by a method such that the determination of the corners of the square is to be made by means of the above rules (1), (2), (3), each of which may be applied any finite number of times. In other words, each new point successively determined in the process of construction is to be obtained as the intersection of two straight lines already determined, or as an intersection of a straight line and a circle already determined, or as an intersection of two circles already determined. A

* See for example the *Mathematical Gazette* for March 1913, where I have treated this point in detail in the Presidential Address to the Mathematical Association.

similar statement applies to the equivalent problem of the rectification of the circle.

This mode of determination of the required figure we may speak of shortly as a Euclidean determination.

Corresponding to any problem of Euclidean determination there is a practical problem of physical Geometry to be carried out by actual construction of straight lines and circles by the use of ruler and compasses. Whenever an ideal problem is soluble as one of Euclidean determination the corresponding practical problem is also a feasible one. The ideal problem has then a solution which is ideally perfect; the practical problem has a solution which is an approximation limited only by the imperfections of the instruments used, the ruler and the compass; and this approximation may be so great that there is no perceptible defect in the result. But it is an error which accounts I think, in large measure, for the aberrations of the circle squarer and the trisector of angles, to assume the converse that, when a practical problem is soluble by the use of the instruments in such a way that the error is negligible or imperceptible, the corresponding ideal problem is also soluble. This is very far from being necessarily the case. It may happen that in the case of a particular ideal problem no solution is obtainable by a finite number of successive Euclidean determinations, and yet that such a finite set gives an approximation to the solution which may be made as close as we please by taking the process far enough. In this case, although the ideal problem is insoluble by the means which are permitted, the practical problem is soluble in the sense that a solution may be obtained in which the error is negligible or imperceptible, whatever standard of possible perceptions we may employ. As we have seen, a Euclidean problem of construction is reducible to the determination of one or more points which satisfy prescribed conditions. Let P be one such point; then it may be possible to determine in Euclidean fashion each point of a set $P_1, P_2, \ldots P_n, \ldots$ of points which converge to P as limiting point, and yet the point P may be incapable of determination by Euclidean procedure. This is what we now know to be the state of things in the case of our special problem of the quadrature of the circle by Euclidean determination. As an ideal problem it is not capable of solution, but the corresponding practical problem is capable of solution with an accuracy bounded only by the limitations of our perceptions and the imperfections of the instruments employed. Ideally we can actually determine by Euclidean methods a square of which the area differs from that of a given circle by less than an arbitrarily prescribed magnitude, although we cannot pass to the limit. We can obtain solutions of the corresponding physical problem which leave nothing to be desired from the practical point of view. Such is the answer which has been obtained to the question raised in this celebrated historical problem of Geometry. I propose to consider in some detail the various modes in which the problem has been attacked by people of various races, and through many centuries; how the modes of attack have been modified by the progressive development of Mathematical tools, and how the final answer, the nature of which had been long anticipated by all competent Mathematicians, was at last found and placed on a firm basis.

General survey of the history of the problem

The history of our problem is typical as exhibiting in a remarkable degree many of the phenomena that are characteristic of the history of Mathematical Science in general. We notice the early attempts at an empirical solution of the problem conceived in a vague and sometimes confused manner; the gradual transition to a clearer notion of the problem as one to be solved subject to precise conditions. We observe also the intimate relation which the mode of regarding the problem in any age had with the state then reached by Mathematical Science in its wider aspect; the essential dependence of the mode of treatment of the problem on the powers of the existing tools. We observe the fact that, as in Mathematics in general, the really great advances, embodying new ideas of far-reaching fruitfulness, have been due to an exceedingly small number of great men; and how a great advance has often been followed by a period in which only comparatively small improvements in, and detailed developments of, the new ideas have been accomplished by a series of men of lesser rank. We observe that there have been periods when for a long series of centuries no advance was made; when the results obtained in a more enlightened age have been forgotten. We observe the times of revival, when the older learning has been rediscovered, and when the results of the progress made in distant countries have been made available as the starting points of new efforts and of a fresh period of activity.

The history of our problem falls into three periods marked out by fundamentally distinct differences in respect of method, of immediate aims, and of equipment in the possession of intellectual tools. The first period embraces the time between the first records of empirical determinations of the ratio of the circumference to the diameter of a circle until the invention of the Differential and Integral Calculus, in the middle of the seventeenth century. This period, in which the ideal of an exact construction was never entirely lost sight of, and was occasionally supposed to have been attained, was the geometrical period, in which the main activity consisted in the approximate determination of π by calculation of the sides or areas of regular polygons in- and circum-scribed to the circle. The theoretical groundwork of the method was the Greek method of Exhaustions. In the earlier part of the period the work of approximation was much hampered by the backward condition of arithmetic due to the fact that our present system of numerical notation had not yet been invented; but the closeness of the approximations obtained in spite of this great obstacle are truly surprising. In the later part of this first period methods were devised by which approximations to the value of π were obtained which required only a fraction of the labour involved in the earlier calculations. At the end of the period the method was developed to so high a degree of perfection that no further advance could be hoped for on the lines laid down by the Greek Mathematicians; for further progress more powerful methods were requisite.

The second period, which commenced in the middle of the seventeenth century, and lasted for about a century, was characterized by the application of the powerful analytical methods provided by the new Analysis to the determination of analytical expressions for the number π in the form of convergent series, products, and continued fractions. The older geometrical forms of investigation gave way to analytical processes in which the functional relationship as applied to the trigonometrical functions became prominent. The new methods of systematic representation gave rise to a race of calculators of π, who, in their consciousness of the vastly enhanced means of calculation placed in their hands by the new Analysis, proceeded to apply the formulae to obtain numerical approximations to π to ever larger numbers of places of decimals, although their efforts were quite useless for the purpose of throwing light upon the true nature of that number. At the end of this period no knowledge had been obtained as regards the number π of a kind likely to throw light upon the possibility or impossibility of the old historical problem of the ideal construction; it was not even definitely known whether the number is rational or irrational. However, one great discovery, destined to furnish the clue to the solution of the problem, was made at this time; that of the relation between the two numbers π and e, as a particular case of those exponential expressions for the trigonometrical functions which form one of the most fundamentally important of the analytical weapons forged during this period.

In the third period, which lasted from the middle of the eighteenth century until late in the nineteenth century, attention was turned to critical investigations of the true nature of the number π itself, considered independently of mere analytical representations. The number was first studied in respect of its rationality or irrationality, and it was shewn to be really irrational. When the discovery was made of the fundamental distinction between algebraic and transcendental numbers, *i.e.* between those numbers which can be, and those numbers which cannot be, roots of an algebraical equation with rational coefficients, the question arose to which of these categories the number π belongs. It was finally established by a method which involved the use of some of the most modern devices of analytical investigation that the number π is transcendental. When this result was combined with the results of a critical investigation of the possibilities of a Euclidean determination, the inference could be made that the number π, being transcendental, does not admit of construction either by a Euclidean determination, or even by a determination in which the use of other algebraic curves besides the straight line and the circle is permitted. The answer to the original question thus obtained is of a conclusively negative character; but it is one in which a clear account is given of the fundamental reasons upon which that negative answer rests.

We have here a record of human effort persisting throughout the best part of four thousand years, in which the goal to be attained was seldom wholly lost sight of. When we look back, in the light of the completed history of the problem, we are able to appreciate the difficulties which in each age restricted the progress which could be made within limits which could not be surpassed by the means then available; we see how, when new weapons became available, a new race of thinkers turned to the further consideration of the problem with a new outlook.

The quality of the human mind, considered in its collective aspect, which most strikes us, in surveying this record, is its colossal patience.

CHAPTER II

THE FIRST PERIOD

Earliest traces of the problem

THE earliest traces of a determination of π are to be found in the Papyrus *Rhind* which is preserved in the British Museum and was translated and explained* by Eisenlohr. It was copied by a clerk, named Ahmes, of the king Raaus, probably about 1700 B.C., and contains an account of older Egyptian writings on Mathematics. It is there stated that the area of a circle is equal to that of a square whose side is the diameter diminished by one ninth; thus $A = (\frac{8}{9})^2 d^2$, or comparing with the formula $A = \frac{1}{4}\pi d^2$, this would give

$$\pi = \frac{256}{81} = 3{\cdot}1604....$$

No account is given of the means by which this, the earliest determination of π, was obtained; but it was probably found empirically.

The approximation $\pi = 3$, less accurate than the Egyptian one, was known to the Babylonians, and was probably connected with their discovery that a regular hexagon inscribed in a circle has its side equal to the radius, and with the division of the circumference into $6 \times 60 \equiv 360$ equal parts.

This assumption ($\pi = 3$) was current for many centuries; it is implied in the Old Testament, 1 Kings vii. 23, and in 2 Chronicles iv. 2, where the following statement occurs:

"Also he made a molten sea of ten cubits from brim to brim, round in compass, and five cubits the height thereof; and a line of thirty cubits did compass it round about."

The same assumption is to be found in the Talmud, where the statement is made "that which in circumference is three hands broad is one hand broad."

* Eisenlohr, *Ein mathematisches Handbuch der alten Ägypter* (Leipzig, 1877).

The earlier Greek Mathematicians

It is to the Greek Mathematicians, the originators of Geometry as an abstract Science, that we owe the first systematic treatment of the problems of the quadrature and rectification of the circle. The oldest of the Greek Mathematicians, Thales of Miletus (640—548 B.C.) and Pythagoras of Samos (580—500 B.C.), probably introduced the Egyptian Geometry to the Greeks, but it is not known whether they dealt with the quadrature of the circle. According to Plutarch (in *De exilio*), Anaxagoras of Clazomene (500—428 B.C.) employed his time during an incarceration in prison on Mathematical speculations, and constructed the quadrature of the circle. He probably made an approximate construction of an equal square, and was of opinion that he had obtained an exact solution. At all events, from this time the problem received continuous consideration.

About the year 420 B.C. Hippias of Elis invented a curve known as the τετραγωνίζουσα or Quadratrix, which is usually connected with the name of Dinostratus (second half of the fourth century) who studied the curve carefully, and who shewed that the use of the curve gives a construction for π.

This curve may be described as follows, using modern notation.

Let a point Q starting at A describe the circular quadrant AB with uniform velocity, and let a point R starting at O describe the radius OB with uniform velocity, and so that if Q and R start simultaneously they will reach the point B simultaneously. Let the point P be the intersection of OQ with a line perpendicular to OB drawn from R. The locus of P is the quadratrix. Letting $\angle QOA = \theta$, and $OR = y$, the ratio y/θ is constant, and equal to $2a/\pi$, where a denotes the radius of the circle. We have

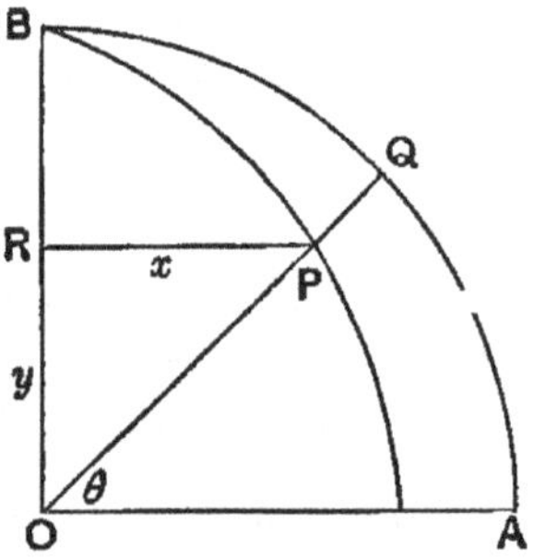

Fig. 1.

$$x = y \cot \theta, \text{ or } x = y \cot \frac{\pi y}{2a},$$

the equation of the curve in rectangular coordinates. The curve will intersect the x axis at the point

$$x = \lim_{y=0} \left(y \cot \frac{\pi y}{2a} \right) = 2a/\pi.$$

If the curve could be constructed, we should have a construction for the length $2a/\pi$, and thence one for π. It was at once seen that the construction of the curve itself involves the same difficulty as that of π.

The problem was considered by some of the Sophists, who made futile attempts to connect it with the discovery of "cyclical square numbers," *i.e.* such square numbers as end with the same cipher as the number itself, as for example $25 = 5^2$, $36 = 6^2$; but the right path to a real treatment of the problem was discovered by Antiphon and further developed by Bryson, both of them contemporaries of Socrates (469—399 B.C.). Antiphon inscribed a square in the circle and passed on to an octagon, 16agon, &c., and thought that by proceeding far enough a polygon would be obtained of which the sides would be so small that they would coincide with the circle. Since a square can always be described so as to be equal to a rectilineal polygon, and a circle can be replaced by a polygon of equal area, the quadrature of the circle would be performed. That this procedure would give only an approximate solution he overlooked. The important improvement was introduced by Bryson of considering circumscribed as well as inscribed polygons; in this procedure he foreshadowed the notion of upper and lower limits in a limiting process. He thought that the area of the circle could be found by taking the mean of the areas of corresponding in- and circum-scribed polygons.

Hippocrates of Chios who lived in Athens in the second half of the fifth century B.C., and wrote the first text book on Geometry, was the first to give examples of curvilinear areas which admit of exact quadrature. These figures are the menisci or lunulae of Hippocrates.

If on the sides of a right-angled triangle ACB semi-circles are described on the same side, the sum of the areas of the two lunes AEC, BDC is equal to that of the triangle ACB. If the right-angled triangle is isosceles, the two lunes are equal, and each of them is half the area of the triangle. Thus the area of a lunula is found.

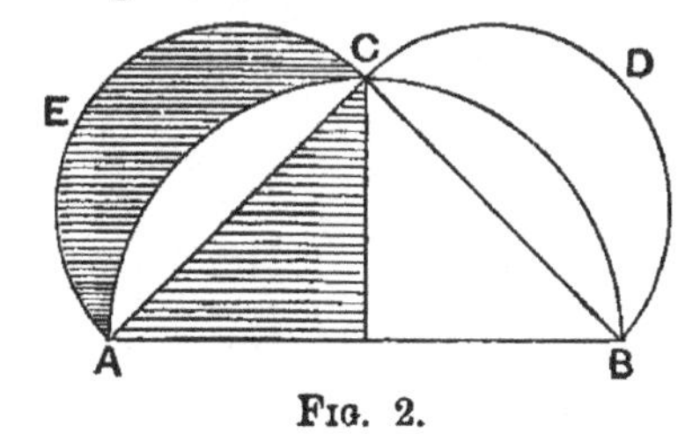

FIG. 2.

If $AC = CD = DB =$ radius OA (see Fig. 3), the semi-circle ACE is $\frac{1}{4}$ of the semi-circle $ACDB$. We have now

$$\text{D}AB - 3\text{D}AC = ACDB - 3 \,.\, \text{meniscus } ACE,$$

and each of these expressions is $\frac{1}{4}\text{D}AB$ or half the circle on $\frac{1}{2}AB$ as diameter. If then the meniscus AEC were quadrable so also would be the circle on $\frac{1}{2}AB$ as diameter. Hippocrates recognized the fact that the meniscus is not quadrable, and he made attempts to find other quadrable lunulae in order to make the quadrature of the circle depend on that of such quadrable lunulae. The question of the existence of various kinds of quadrable lunulae was taken up by Th. Clausen* in 1840, who discovered four other quadrable lunulae in addition to the one mentioned above. The question was considered in a general manner by Professor Landan† of Göttingen in 1890, who pointed out that two of the four lunulae which Clausen supposed to be new were already known to Hippocrates.

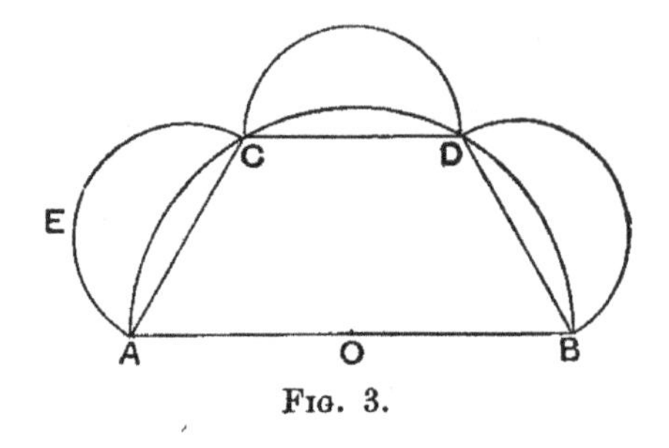

Fig. 3.

From the time of Plato (429—348 B.C.), who emphasized the distinction between Geometry which deals with incorporeal things or images of pure thought and Mechanics which is concerned with things in the external world, the idea became prevalent that problems such as that with which we are concerned should be solved by Euclidean determination only, equivalent on the practical side to the use of two instruments only, the ruler and the compass.

The work of Archimedes

The first really scientific treatment of the problem was undertaken by the greatest of all the Mathematicians of antiquity, Archimedes (287—212 B.C.). In order to understand the mode in which he actually established his very important approximation to the value of π it is necessary for us to consider in some detail the Greek method of dealing with problems of limits, which in the hands of Archimedes provided a method of performing genuine integrations, such as his determination of the area of a segment of a parabola, and of a considerable number of areas and volumes.

This method is that known as the method of exhaustions, and rests on a principle stated in the enunciation of Euclid x. 1, as follows:

"Two unequal magnitudes being set out, if from the greater there be subtracted a magnitude greater than its half, and from that which

* *Journal für Mathematik*, vol. 21, p. 375.

† *Archiv Math. Physik* (3) 4 (1903).

is left a magnitude greater than its half, and if this process be repeated continually, there will be left some magnitude which will be less than the lesser magnitude set out."

This principle is deduced by Euclid from the axiom that, if there are two magnitudes of the same kind, then a multiple of the smaller one can be found which will exceed the greater one. This latter axiom is given by Euclid in the form of a definition of ratio (Book v. def. 4), and is now known as the axiom of Archimedes, although, as Archimedes himself states in the introduction to his work on the quadrature of the parabola, it was known and had been already employed by earlier Geometers. The importance of this so-called axiom of Archimedes, now generally considered as a postulate, has been widely recognized in connection with the modern views as to the arithmetic continuum and the theory of continuous magnitude. The attention of Mathematicians was directed to it by O. Stolz*, who shewed that it was a consequence of Dedekind's postulate relating to "sections." The possibility of dealing with systems of numbers or of magnitudes for which the principle does not hold has been considered by Veronese and other Mathematicians, who contemplate non-Archimedean systems, *i.e.* systems for which this postulate does not hold. The acceptance of the postulate is equivalent to the ruling out of infinite and of infinitesimal magnitudes or numbers as existent in any system of magnitudes or of numbers for which the truth of the postulate is accepted.

The example of the use of the method of exhaustions which is most familiar to us is contained in the proof given in Euclid xii. 2, that the areas of two circles are to one another as the squares on their diameters. This theorem which is a presupposition of the reduction of the problem of squaring the circle to that of the determination of a definite ratio π is said to have been proved by Hippocrates, and the proof given by Euclid is pretty certainly due to Eudoxus, to whom various other applications of the method of Exhaustions are specifically attributed by Archimedes. Euclid shews that the circle can be "exhausted" by the inscription of a sequence of regular polygons each of which has twice as many sides as the preceding one. He shews that the area of the inscribed square exceeds half the area of the circle; he then passes to an octagon by bisecting the arcs bounded by the sides of the square. He shews that the excess of the area of the circle over that of the octagon is less than half what is left of the circle when the square is removed from it, and so on through the further stages of the process.

* See *Math. Annalen*, vol. 22, p. 504, and vol. 39, p. 107.

The truth of the theorem is then inferred by shewing that a contrary assumption leads to a contradiction.

A study of the works of Archimedes, now rendered easily accessible to us in Sir T. L. Heath's critical edition, is of the greatest interest not merely from the historical point of view but also as affording a very instructive methodological study of rigorous treatment of problems of determination of limits. The method by which Archimedes and other Greek Mathematicians contemplated limit problems impresses one, apart from the geometrical form, with its essentially modern way of regarding such problems. In the application of the method of exhaustions and its extensions no use is made of the ideas of the infinite or the infinitesimal; there is no jumping to the limit as the supposed end of an essentially endless process, to be reached by some inscrutable *saltus*. This passage to the limit is always evaded by substituting a proof in the form of a *reductio ad absurdum*, involving the use of inequalities such as we have in recent times again adopted as appropriate to a rigorous treatment of such matters. Thus the Greeks, who were however thoroughly familiar with all the difficulties as to infinite divisibility, continuity, &c., in their mathematical proofs of limit theorems never involved themselves in the morass of indivisibles, indiscernibles, infinitesimals, &c., in which the Calculus after its invention by Newton and Leibnitz became involved, and from which our own text books are not yet completely free.

The essential rigour of the processes employed by Archimedes, with such fruitful results, leaves, according to our modern views, one point open to criticism. The Greeks never doubted that a circle has a definite area in the same sense that a rectangle has one; nor did they doubt that a circle has a length in the same sense that a straight line has one. They had not contemplated the notion of non-rectifiable curves, or non-quadrable areas; to them the existence of areas and lengths as definite magnitudes was obvious from intuition. At the present time we take only the length of a segment of a straight line, the area of a rectangle, and the volume of a rectangular parallelepiped as primary notions, and other lengths, areas, and volumes we regard as derivative, the actual existence of which in accordance with certain definitions requires to be established in each individual case or in particular classes of cases. For example, the measure of the length of a circle is defined thus: A sequence of inscribed polygons is taken so that the number of sides increases indefinitely as the sequence proceeds, and such that the length of the greatest side of the polygon diminishes indefinitely, then if the numbers which represent the perimeters of the successive polygons form a convergent sequence, of which the arithmetical limit is one and the same number for all sequences of polygons which satisfy the prescribed conditions, the circle has a length represented by this limit. It must be proved that this limit exists and is independent of the particular sequence employed, before we are entitled to regard the circle as rectifiable.

In his work κύκλου μέτρησις, the measurement of a circle, Archimedes proves the following three theorems.

(1) The area of any circle is equal to a right-angled triangle in which one of the sides about the right angle is equal to the radius, and the other to the circumference, of the circle.

(2) The area of the circle is to the square on its diameter as 11 to 14.

(3) The ratio of the circumference of any circle to its diameter is less than $3\frac{1}{7}$ but greater than $3\frac{10}{71}$.

It is clear that (2) must be regarded as entirely subordinate to (3). In order to estimate the accuracy of the statement in (3), we observe that

$$3\tfrac{1}{7} = 3{\cdot}14285\ldots,\quad 3\tfrac{10}{71} = 3{\cdot}14084\ldots,\quad \pi = 3{\cdot}14159\ldots.$$

In order to form some idea of the wonderful power displayed by Archimedes in obtaining these results with the very limited means at his disposal, it is necessary to describe briefly the details of the method he employed.

His first theorem is established by using sequences of in- and circum-scribed polygons and a *reductio ad absurdum*, as in Euclid XII. 2, by the method already referred to above.

In order to establish the first part of (3), Archimedes considers a regular hexagon circumscribed to the circle.

In the figure, AC is half one of the sides of this hexagon. Then

$$\frac{OA}{AC} = \sqrt{3} > \frac{265}{153}.$$

Bisecting the angle AOC, we obtain AD half the side of a regular circumscribed 12agon. It is then shewn that $\dfrac{OD}{DA} > \dfrac{591\frac{1}{8}}{153}$. If OE is the bisector of the angle DOA, AE is half the side of a circumscribed 24agon, and it is then shewn that $\dfrac{OE}{EA} > \dfrac{1172\frac{1}{8}}{153}$. Next, bisecting EOA, we obtain AF the half side of a 48agon, and it is shewn that $\frac{OF}{FA} > \frac{2339\frac{1}{4}}{153}$. Lastly if OG (not shewn in the figure) be the bisector of FOA, AG is the half side of a regular 96agon circumscribing the circle, and it is shewn that $\frac{OA}{AG} > \frac{4673\frac{1}{2}}{153}$, and thence that the ratio of the diameter to the perimeter of the 96agon

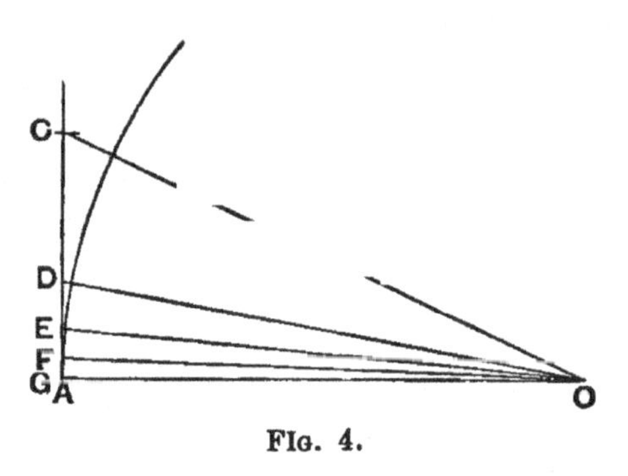

FIG. 4.

is $> \frac{4673\frac{1}{2}}{14688}$, and it is deduced that the circumference of the circle, which is less than the perimeter of the polygon, is $< 3\frac{1}{7}$ of the diameter. The second part of the theorem is obtained in a similar manner by determination of the side of a regular 96agon inscribed in the circle.

In the course of his work, Archimedes assumes and employs, without explanation as to how the approximations were obtained, the following estimates of the values of square roots of numbers:

$$\tfrac{1351}{780} > \sqrt{3} > \tfrac{265}{153}, \quad 3013\tfrac{3}{4} > \sqrt{9082321}, \quad 1838\tfrac{9}{11} > \sqrt{3380929},$$

$$1009\tfrac{1}{6} > \sqrt{1018405}, \quad 2017\tfrac{1}{4} > \sqrt{4069284\tfrac{1}{36}}, \quad 591\tfrac{1}{8} < \sqrt{349450},$$

$$1172\tfrac{1}{8} < \sqrt{1373943\tfrac{33}{64}}, \quad 2339\tfrac{1}{4} < \sqrt{5472132\tfrac{1}{16}}.$$

In order to appreciate the nature of the difficulties in the way of obtaining these approximations we must remember the backward condition of Arithmetic with the Greeks, owing to the fact that they possessed a system of notation which was exceedingly inconvenient for the purpose of performing arithmetical calculations.

The letters of the alphabet together with three additional signs were employed, each letter being provided with an accent or with a short horizontal stroke; thus the nine integers

1, 2, 3, 4, 5, 6, 7, 8, 9 were denoted by α′, β′, γ′, δ′, ε′, ϛ′, ζ′, η′, θ′,

the multiples of 10,

10, 20, 30, ... 90 were denoted by ι′, κ′, λ′, μ′, ν′, ξ′, ο′, π′, ϙ′, the multiples of 100,

100, 200, ... 900 by ρ', σ', τ', υ', φ', χ', ψ', ω', ϡ'.

The intermediate numbers were expressed by juxtaposition, representing here addition, the largest number being placed on the left, the next largest following, and so on in order. There was no sign for zero. Thousands were represented by the same letters as the first nine integers but with a small dash in front and below the line; thus for example ͵δ' was 4000, and 1913 was expressed by ͵αϡιγ' or $\overline{͵\alpha ϡ \iota\gamma}$. 10000 and higher numbers were expressed by using the ordinary numerals with M or Mυ as an abbreviation for the word *μυριάς*; the number of myriads, or the multiple of 10000, was generally written over the abbreviation, thus 349450 was $\overset{\lambda\delta}{\mathrm{M}}$͵θων'. A variety of devices were employed for the representation of fractions*

The determinations of square roots such as $\sqrt{3}$ by Archimedes were much closer than those of earlier Greek writers. There has been much speculation as to the method he must have employed in their determination. There is reason to believe that he was acquainted with the method of approximation that we should denote by

$$a \pm \frac{b}{2a} > \sqrt{a^2 \pm b} > a \pm \frac{b}{2a \pm 1}.$$

Various alternative explanations have been suggested; some of these suggest that a method equivalent to the use of approximation by continued fractions was employed.

A full discussion of this matter will be found in Sir T. L. Heath's work on Archimedes.

The treatise of Archimedes on the measurement of the circle must be regarded as the one really great step made by the Greeks towards the solution of the problem; in fact no essentially new mode of attack was made until the invention of the Calculus provided Mathematicians with new weapons. In a later writing which has been lost, but which is mentioned by Hero, Archimedes found a still closer approximation to π.

The essential points of the method of Archimedes, when generalized and expressed in modern notation, consist of the following theorems:

(1) The inequalities $\sin\theta < \theta < \tan\theta$.

* For an interesting account of the Arithmetic of Archimedes, see Heath's *Works of Archimedes*, Chapter IV.

(2) The relations for the successive calculation of the perimeters and areas of polygons inscribed and circumscribed to a circle.

Denoting by p_n, a_n the perimeter and area of an inscribed regular polygon of n sides, and by P_n, A_n the perimeter and area of a circumscribed regular polygon of n sides, these relations are

$$p_{2n} = \sqrt{p_n P_{2n}}, \quad a_{2n} = \sqrt{a_n A_n},$$

$$P_{2n} = \frac{2p_n P_n}{p_n + P_n}, \quad A_{2n} = \frac{2a_{2n} A_n}{a_{2n} + A_n}.$$

Thus the two series of magnitudes

$$P_n, p_n, P_{2n}, p_{2n}, P_{4n}, p_{4n}, \ldots,$$

$$A_n, a_{2n}, A_{2n}, a_{4n}, A_{4n}, a_{8n}, \ldots,$$

are calculated successively in accordance with the same law. In each case any element is calculated from the two preceding ones by taking alternately their harmonic and geometric means. This system of formulae is known as the Archimedean Algorithm; by means of it the chords and tangents of the angles at the centre of such polygons as are constructible can be calculated. By methods essentially equivalent to the use of this algorithm the sines and tangents of small angles were obtained to a tolerably close approximation. For example, Aristarchus (250 B.C.) obtained the limits $\frac{1}{45}$ and $\frac{1}{60}$ for sin 1°.

The work of the later Greeks

Among the later Greeks, Hipparchus (180—125 B.C.) calculated the first table of chords of a circle and thus founded the science of Trigonometry. But the greatest step in this direction was made by Ptolemy (87—165 A.D.) who calculated a table of chords in which the chords of all angles at intervals of $\frac{1}{2}$° from 0° to 180° are contained, and thus constructed a trigonometry that was not surpassed for 1000 years. He was the first to obtain an approximation to π more exact than that of Archimedes; this was expressed in sexagesimal measure by 3° 8′ 30″ which is equivalent to

$$3 + \frac{8}{60} + \frac{30}{3600} \text{ or } 3\tfrac{17}{120} \equiv 3{\cdot}14166\ldots.$$

The work of the Indians

We have now to pass over to the Indian Mathematicians. Áryabhatta (about 500 A.D.) knew the value

$$\frac{62832}{20000} \equiv 3{\cdot}1416 \text{ for } \pi.$$

The same value in the form $\frac{3927}{1250}$ was given by Bhâskara (born 1114 A.D.)

in his work *The crowning of the system*; and he describes this value as exact, in contrast with the inexact value $\frac{22}{7}$. His commentator Gancea explains that this result was obtained by calculating the perimeters of polygons of 12, 24, 48, 96, 192, and 384 sides, by the use of the formula

$$a_{2n} = \sqrt{2 - \sqrt{4 - a_n^2}}$$

connecting the sides of inscribed polygons of $2n$ and n sides respectively, the radius being taken as unity. If the diameter is 100, the side of an inscribed 384agon is $\sqrt{98694}$ which leads to the above value* given by Áryabhatta. Brahmagupta (born 598 A.D.) gave as the exact value $\pi = \sqrt{10}$. Hankel has suggested that this was obtained as the supposed limit ($\sqrt{1000}$) of $\sqrt{965}$, $\sqrt{981}$, $\sqrt{986}$, $\sqrt{987}$ (diameter 10), the perimeters of polygons of 12, 24, 48, 96 sides, but this explanation is doubtful. It has also been suggested that it was obtained by the approximate formula

$$\sqrt{a^2 + x} = a + \frac{1}{2a + x},$$

which gives $\sqrt{10} = 3 + \frac{1}{7}$.

The work of the Chinese Mathematicians

The earliest Chinese Mathematicians, from the time of Chou-Kong who lived in the 12th century B.C., employed the approximation $\pi = 3$. Some of those who used this approximation were mathematicians of considerable attainments in other respects.

According to the Sui-shu, or *Records of the Sui dynasty*, there were a large number of circle-squarers, who calculated the length of the circular circumference, obtaining however divergent results.

Chang Hing, who died in 139 A.D., gave the rule

$$\frac{(\text{circumference})^2}{(\text{perimeter of circumscribed square})^2} = \frac{5}{8},$$

which is equivalent to $\pi = \sqrt{10}$.

Wang Fau made the statement that if the circumference of a circle is 142 the diameter is 45; this is equivalent to $\pi = 3{\cdot}1555\ldots$. No record has been found of the method by which this result was obtained.

* See Colebrooke's *Algebra with arithmetic and mensuration, from the Sanscrit of Brahmagupta and Bhâskara*, London, 1817.

Liu Hui published in 263 A.D. an *Arithmetic in nine sections* which contains a determination of π. Starting with an inscribed regular hexagon, he proceeds to the inscribed dodecagon, 24agon, and so on, and finds the ratio of the circumference to the diameter to be 157 : 50, which is equivalent to $\pi = 3{\cdot}14$.

By far the most interesting Chinese determination was that of the great Astronomer *Tsu Ch'ung-chih* (born 430 A.D.). He found the two values $\frac{22}{7}$ and $\frac{355}{113}$ ($= 3{\cdot}1415929$ ..). In fact he proved that 10π lies between $31{\cdot}415927$ and $31{\cdot}415926$, and deduced the value $\frac{355}{113}$.

The value $\frac{22}{7}$ which is that of Archimedes he spoke of as the "inaccurate" value, and $\frac{355}{113}$ as the "accurate value." This latter value was not obtained either by the Greeks or the Hindoos, and was only rediscovered in Europe more than a thousand years later, by Adriaen Anthonisz. The later Chinese Mathematicians employed for the most part the "inaccurate" value, but the "accurate" value was rediscovered by Chang Yu-chin, who employed an inscribed polygon with 2^{14} sides.

The work of the Arabs

In the middle ages a knowledge of Greek and Indian mathematics was introduced into Europe by the Arabs, largely by means of Arabic translations of Euclid's elements, Ptolemy's σύνταξις, and treatises by Appollonius and Archimedes, including the treatise of Archimedes on the measurement of the circle.

The first Arabic Mathematician Muhammed ibn Mûsâ Alchwarizmi, at the beginning of the ninth century, gave the Greek value $\pi = 3\frac{1}{7}$, and the Indian values $\pi = \sqrt{10}$, $\pi = \frac{62832}{20000}$, which he states to be of Indian origin. He introduced the Indian system of numerals which was spread in Europe at the beginning of the 13th century by Leonardo Pisano, called Fibonacci.

The time of the Renaissance

The greatest Christian Mathematician of medieval times, Leonardo Pisano (born at Pisa at the end of the 12th century), wrote a work entitled *Practica geometriae*, in 1220, in which he improved on the results of Archimedes, using the same method of employing the in- and circum-scribed 96agons. His limits are $\frac{1440}{458\frac{1}{5}} = 3{\cdot}1427$ and

$\frac{1440}{458\frac{4}{9}} = 3{\cdot}1410\ldots$, whereas $3\frac{1}{7} = 3{\cdot}1428$, $3\frac{10}{71} = 3{\cdot}1408\ldots$ were the values given by Archimedes. From these limits he chose $\frac{1440}{458\frac{1}{3}}$ or $\pi = 3{\cdot}1418\ldots$ as the mean result.

During the period of the Renaissance no further progress in the problem was made beyond that due to *Leonardo Pisano*; some later writers still thought that $3\frac{1}{7}$ was the exact value of π. George Purbach (1423—1461), who constructed a new and more exact table of sines of angles at intervals of 10′, was acquainted with the Archimedean and Indian values, which he fully recognized to be approximations only. He expressed doubts as to whether an exact value exists. Cardinal Nicholas of Cusa (1401—1464) obtained $\pi = 3{\cdot}1423$ which he thought to be the exact value. His approximations and methods were criticized by Regiomontanus (Johannes Müller, 1436—1476), a great mathematician who was the first to shew how to calculate the sides of a spherical triangle from the angles, and who calculated extensive tables of sines and tangents, employing for the first time the decimal instead of the sexagesimal notation.

The fifteenth and sixteenth centuries

In the fifteenth and sixteenth centuries great improvements in trigonometry were introduced by Copernicus (1473—1543), Rheticus (1514—1576), Pitiscus (1561—1613), and Johannes Kepler (1571—1630).

These improvements are of importance in relation to our problem, as forming a necessary part of the preparation for the analytical developments of the second period.

In this period Leonardo da Vinci (1452—1519) and Albrecht Dürer (1471—1528) should be mentioned, on account of their celebrity, as occupying themselves with our subject, without however adding anything to the knowledge of it.

Orontius Finaeus (1494—1555) in a work *De rebus mathematicis hactenus desiratis*, published after his death, gave two theorems which were later established by Huyghens, and employed them to obtain the limits $\frac{22}{7}$, $\frac{245}{78}$ for π; he appears to have asserted that $\frac{245}{78}$ is the exact value. His theorems when generalized are expressed in our notation by the fact that θ is approximately equal to $(\sin^2\theta\tan\theta)^{\frac{1}{3}}$.

The development of the theory of equations which later became of fundamental importance in relation to our problem was due to the Italian Mathematicians of the 16th century, Tartaglia (1506—1559), Cardano (1501—1576), and Ferrari (1522—1565).

The first to obtain a more exact value of π than those hitherto known in Europe was Adriaen Anthonisz (1527—1607) who rediscovered the Chinese value $\pi = \frac{355}{113} = 3{\cdot}1415929\ldots$, which is correct to 6 decimal places. His son Adriaen who took the name of Metius (1571—1635), published this value in 1625, and explained that his father had obtained the approximations $\frac{333}{106} < \pi < \frac{377}{120}$ by the method of Archimedes, and had then taken the mean of the numerators and denominators, thus obtaining his value.

The first explicit expression for π by an infinite sequence of operations was obtained by Vieta (François Viète, 1540—1603). He proved that, if two regular polygons are inscribed in a circle, the first having half the number of sides of the second, then the area of the first is to that of the second as the supplementary chord of a side of the first polygon is to the diameter of the circle. Taking a square, an octagon, then polygons of 16, 32, ... sides, he expressed the supplementary chord of the side of each, and thus obtained the ratio of the area of each polygon to that of the next. He found that, if the diameter be taken as unity, the area of the circle is

$$2\,\frac{1}{\sqrt{\frac{1}{2}}\sqrt{\frac{1}{2}+\frac{1}{2}\sqrt{\frac{1}{2}}}\sqrt{\frac{1}{2}+\frac{1}{2}\sqrt{\frac{1}{2}+\frac{1}{2}\sqrt{\frac{1}{2}}}}\cdots},$$

from which we obtain

$$\frac{\pi}{2} = \frac{1}{\sqrt{\frac{1}{2}}\sqrt{\frac{1}{2}+\frac{1}{2}\sqrt{\frac{1}{2}}}\cdots}.$$

It may be observed that this expression is obtainable from the formula

$$\theta = \frac{\sin\theta}{\cos\frac{\theta}{2}\cos\frac{\theta}{4}\cos\frac{\theta}{8}\cdots} \quad (\theta < \pi)$$

afterwards obtained by Euler, by taking $\theta = \frac{\pi}{2}$.

Applying the method of Archimedes, starting with a hexagon and proceeding to a polygon of $2^{16}.6$ sides, Vieta shewed that, if the diameter of the circle be 100000, the circumference is $> 314159\frac{26535}{100000}$ and is $< 3{\cdot}14159\frac{26537}{100000}$; he thus obtained π correct to 9 places of decimals.

Adrianus Romanus (Adriaen van Rooman, born in Lyons, 1561—1615) by the help of a $15 \cdot 2^{24}$agon calculated π to 15 places of decimals.

Ludolf van Ceulen (Cologne) (1539—1610), after whom the number π is still called in Germany "Ludolph's number," is said to have calculated π to 35 places of decimals. According to his wish the value was engraved on his tombstone which has been lost. In his writing *Van den Cirkel* (Delft, 1596) he explained how, by employing the method of Archimedes, using in- and circum-scribed polygons up to the $60 \cdot 2^{29}$agon, he obtained π to 20 decimal places. Later, in his work *De Arithmetische en Geometrische fondamenten* he obtained the limits given by

$$3 \frac{14159265358979323846264338327950}{100000000000000000000000000000000},$$

and the same expression with 1 instead of 0 in the last place of the numerator.

The work of Snellius and Huyghens

In a work *Cyclometricus*, published in 1621, Willebrod Snellius (1580—1626) shewed how narrower limits can be determined, without increasing the number of sides of the polygons, than in the method of Archimedes. The two theorems, equivalent to the approximations

$$\tfrac{1}{3}(2 \sin\theta + \tan\theta) < \theta < 3/(2 \operatorname{cosec}\theta + \cot\theta),$$

by which he attained this result were not strictly proved by him, and were afterwards established by Huyghens; the approximate formula $\theta = \dfrac{3 \sin\theta}{2 + \cos\theta}$ had been already obtained by Nicholas of Cusa (1401—1464). Using in- and circum-scribed hexagons the limits 3 and 3·464 are obtained by the method of Archimedes, but Snellius obtained from the hexagons the limits 3·14022 and 3·14160, closer than those obtained by Archimedes from the 96agon. With the 96agon he found the limits 3·1415926272 and 3·1415928320. Finally he verified Ludolf's determination with a great saving of labour, obtaining 34 places with the 2^{30}agon, by which Ludolf had only obtained 14 places. Grunberger* calculated 39 places by the help of the formulae of Snellius.

The extreme limit of what can be obtained on the geometrical lines laid down by Archimedes was reached in the work of Christian Huyghens (1629—1665). In his work† *De circuli magnitudine inventa*,

* *Elementa Trigonometriae*, Rome, 1630.

† A study of the German Translation by Rudio will repay the trouble.

which is a model of geometrical reasoning, he undertakes by improved methods to make a careful determination of the area of a circle. He establishes sixteen theorems by geometrical processes, and shews that by means of his theorems three times as many places of decimals can be obtained as by the older method. The determination made by Archimedes he can get from the triangle alone. The hexagon gives him the limits 3·1415926533 and 3·1415926538.

The following are the theorems proved by Huyghens:

I. If ABC is the greatest triangle in a segment less than a semi-circle, then

$$\triangle ABC < 4(\triangle AEB + \triangle BFC),$$

where AEB, BFC are the greatest triangles in the segments AB, BC.

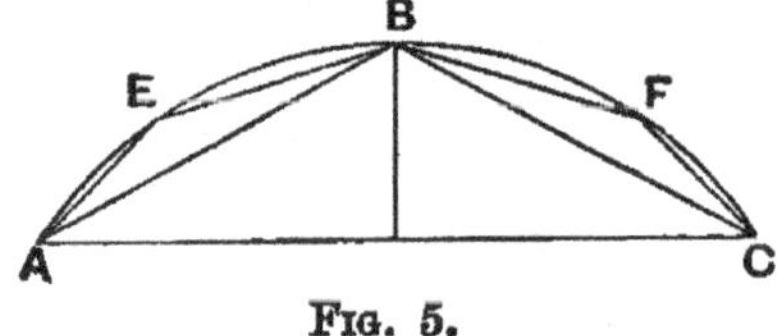

FIG. 5.

II. $\triangle FEG > \frac{1}{2} \triangle ABC$,

where ABC is the greatest triangle in the segment.

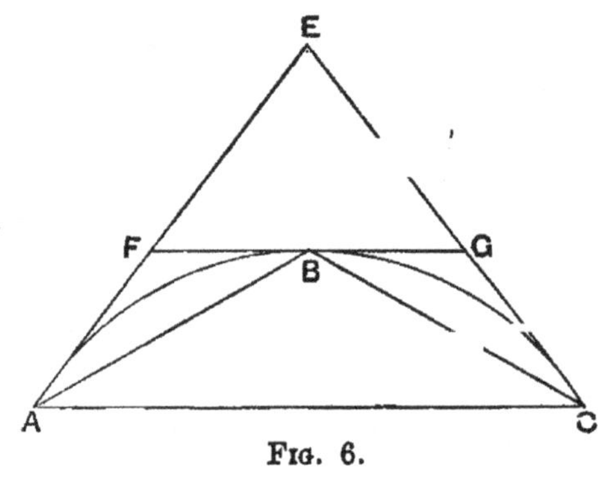

FIG. 6.

III. $$\frac{\text{segment } ACB}{\triangle ACB} > \frac{4}{3},$$

provided the segment is less than the semi-circle.

This theorem had already been given by Hero.

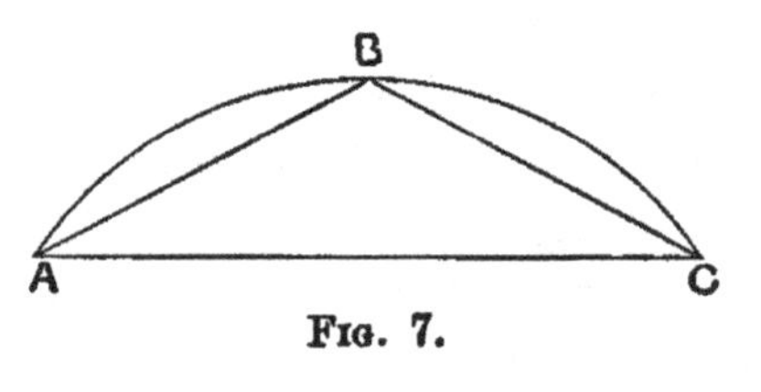

FIG. 7.

IV. $$\frac{\text{segment } ACB}{\triangle ATC} < \frac{2}{3}.$$

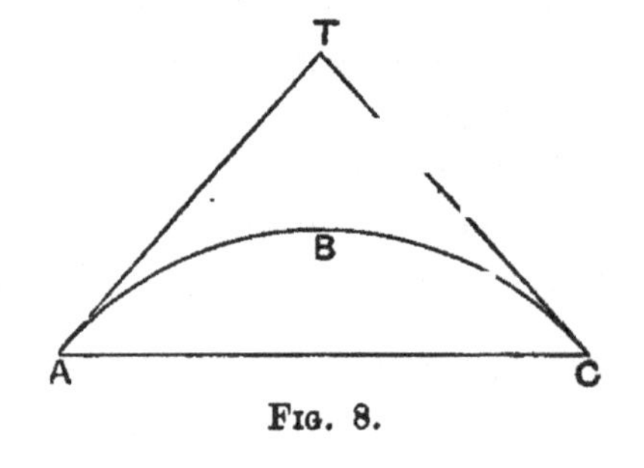

FIG. 8.

V. If A_n is the area of an inscribed regular polygon of n sides, and S the area of the circle, then $S > A_{2n} + \frac{1}{3}(A_{2n} - A_n)$.

VI. If A_n' is the area of the circumscribed regular polygon of n sides, then $S < \frac{2}{3}A_n' + \frac{1}{3}A_n$.

VII. If C_n denotes the perimeter of the inscribed polygon, and C the circumference of the circle, then $C > C_{2n} + \frac{1}{3}(C_{2n} - C_n)$.

VIII. $\frac{2}{3}CD + \frac{1}{3}EF >$ arc CE, where E is any point on the circle.

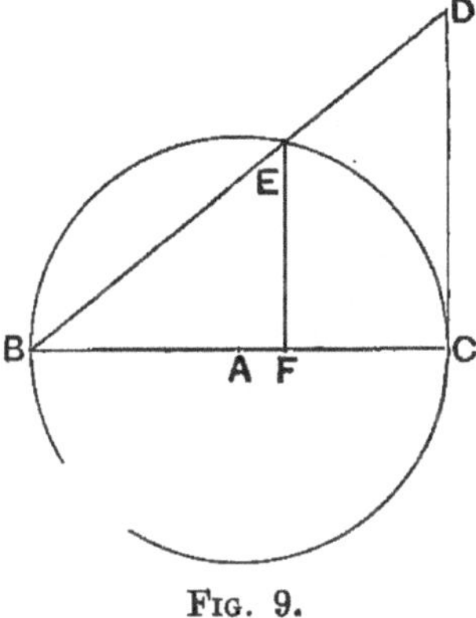

FIG. 9.

IX. $$C < \tfrac{2}{3}C_n + \tfrac{1}{3}C_n',$$ where C_n' is the perimeter of the circumscribed polygon of n sides.

X. If a_n, a_n' denote the sides of the in- and circum-scribed polygons, then $a_{2n}^2 = a_{2n}' \cdot \frac{1}{2}a_n$.

XI. $C <$ the smaller of the two mean proportionals between C_n and C_n'.

$S <$ the similar polygon whose perimeter is the larger of the two mean proportionals.

XII. If ED equals the radius of the circle, then $BG >$ arc BF.

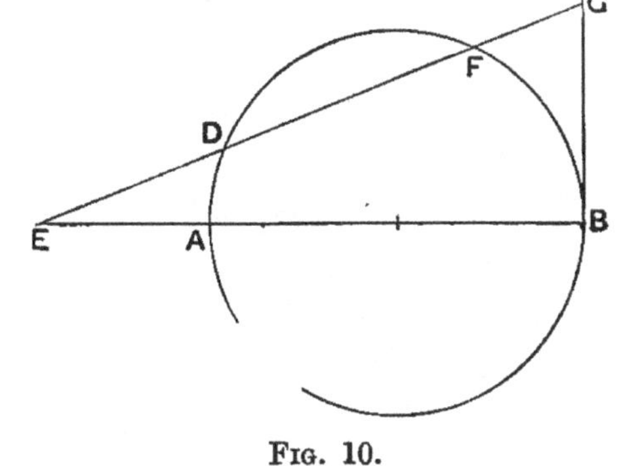

FIG. 10.

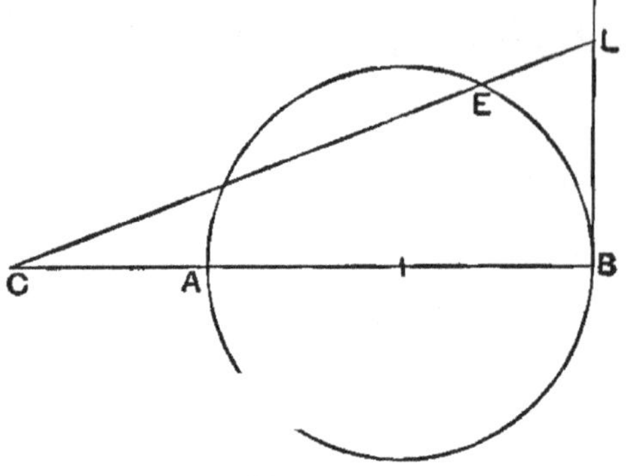

Fig. 11.

XIII. If AC = radius of the circle, then $BL <$ arc BE.

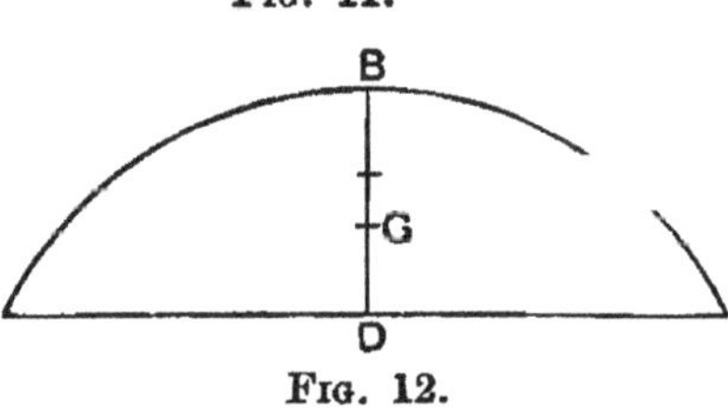

Fig. 12.

XIV. If G is the centroid of the segment, then

$$BG > GD \text{ and } < \tfrac{3}{2}GD.$$

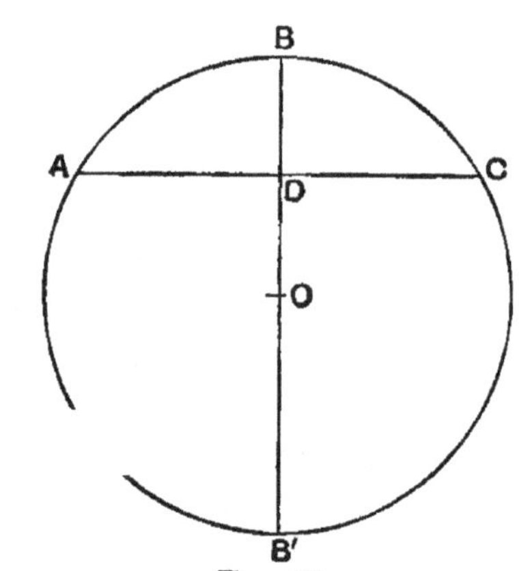

Fig. 13.

XV. $$\frac{\text{segment } ABC}{\triangle ABC} > \frac{4}{3}$$

and $$< 3\tfrac{1}{3} \cdot \frac{B'D}{BB' + 3OD}.$$

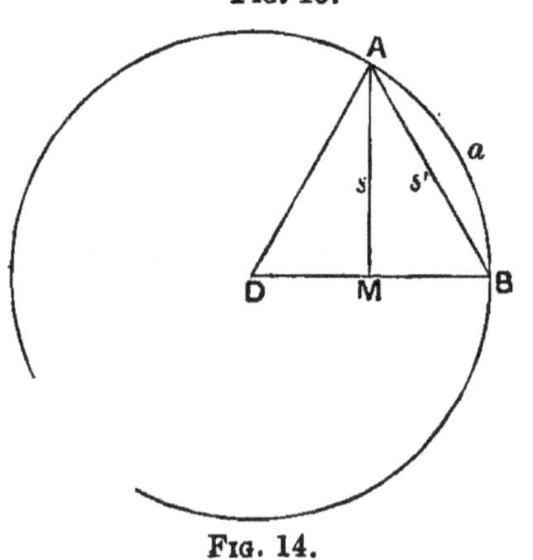

Fig. 14.

XVI. If a denote the arc (< semi-circle), and s, s' its sine and its chord respectively, then

$$s' + \frac{s' - s}{3} < a < s' + \frac{s' - s}{3} \cdot \frac{4s' + s}{2s' + 3s}.$$

This is equivalent, as Huyghens points out, to

$$p_{2n} + \frac{p_{2n} - p_n}{3} < C$$
$$< p_{2n} + \frac{p_{2n} - p_n}{3} \cdot \frac{4p_{2n} + p_n}{2p_{2n} + 3p_n},$$

where p_n is the perimeter of a regular inscribed polygon of n sides, and C is the circumference of the circle.

The work of Gregory

The last Mathematician to be mentioned in connection with the development of the method of Archimedes is James Gregory (1638—1675), Professor in the Universities of St Andrews and Edinburgh, whose important work in connection with the development of the new Analysis we shall have to refer to later. Instead of employing the perimeters of successive polygons, he calculated their areas, using the formulae

$$A_{2n}' = \frac{2A_n A_n'}{A_n + A_{2n}} = \frac{2A_n' A_{2n}}{A_n' + A_{2n}};$$

where A_n, A_n' denote the areas of in- and circum-scribed regular n-agons; he also employed the formula $A_{2n} = \sqrt{A_n A_n'}$ which had been obtained by Snellius. In his work *Exercitationes geometricae* published in 1668, he gave a whole series of formulae for approximations on the lines of Archimedes. But the most interesting step which Gregory took in connection with the problem was his attempt to prove, by means of the Archimedean algorithm, that the quadrature of the circle is impossible. This is contained in his work *Vera circuli et hyperbolae quadratura* which is reprinted in the works of Huyghens (*Opera varia* I, pp. 315—328) who gave a refutation of Gregory's proof. Huyghens expressed his own conviction of the impossibility of the quadrature, and in his controversy with Wallis remarked that it was not even decided whether the area of the circle and the square of the diameter are commensurable or not. In default of a theory of the distinction between algebraic and transcendental numbers, the failure of Gregory's proof was inevitable. Other such attempts were made by Lagny (*Paris Mém.* 1727, p. 124), Saurin (*Paris Mém.* 1720), Newton (*Principia* I, 6, Lemma 28), and Waring (*Proprietates algebraicarum curvarum*) who maintained that no algebraical oval is quadrable. Euler also made some attempts in the same direction (*Considerationes cyclometricae, Novi Comm. Acad. Petrop.* XVI, 1771); he observed that the irrationality of π must first be established, but that this would not of itself be sufficient to prove the impossibility of the quadrature. Even as early as 1544, Michael Stifel, in his *Arithmetic integra*, expressed the opinion that the construction is impossible. He emphasized the distinction between a theoretical and a practical construction.

The work of Descartes

The great Philosopher and Mathematician René Descartes (1596—1650), of immortal fame as the inventor of coordinate geometry, regarded the problem from a new point of view. A given straight line being taken as equal to the circumference of a circle he proposed to determine the diameter by the following construction:

Take AB one quarter of the given straight line. On AB describe the square $ABCD$; by a known process a point C_1 on AC produced, can be so determined that the rectangle $BC_1 = \frac{1}{4}ABCD$. Again C_2 can be so determined that rect. $B_1C_2 = \frac{1}{4}BC_1$; and so on indefinitely. The diameter required is given by AB_ω, where B_ω is the limit to which B, B_1, B_2, ... converge. To see the reason of this, we can shew that AB is the diameter of the circle inscribed in $ABCD$, that AB_1 is the diameter of the circle circumscribed by the regular octagon having the same perimeter as the square; and generally that AB_n is the diameter of the regular 2^{n+2}-agon having the same perimeter as the square. To verify this, let

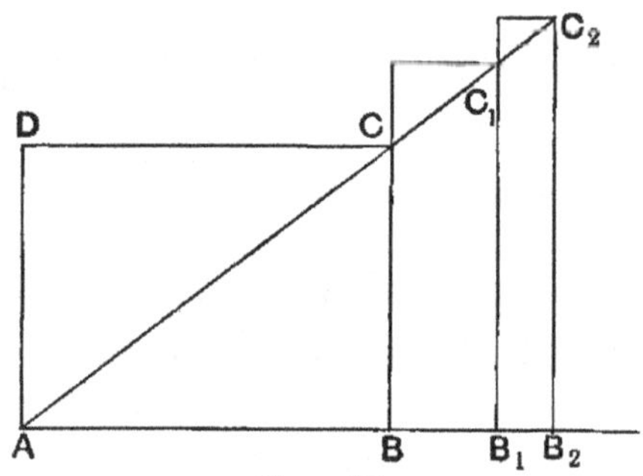

$$x_n = AB_n, \quad x_0 = AB;$$

then by the construction,

$$x_n(x_n - x_{n-1}) = \frac{1}{4^n} x_0^2,$$

FIG. 15.

and this is satisfied by $x_n = \frac{4x_0}{2^n} \cot \frac{\pi}{2^n}$; thus

$$\lim x_n = \frac{4x_0}{\pi} = \text{diameter of the circle.}$$

This process was considered later by Schwab (Gergonne's *Annales de Math.* vol. VI), and is known as the process of isometers.

This method is equivalent to the use of the infinite series

$$\frac{4}{\pi} = \tan\frac{\pi}{4} + \frac{1}{2}\tan\frac{\pi}{8} + \frac{1}{4}\tan\frac{\pi}{16} + \dots,$$

which is a particular case of the formula

$$\frac{1}{x} - \cot x = \frac{1}{2}\tan\frac{x}{2} + \frac{1}{4}\tan\frac{x}{4} + \frac{1}{8}\tan\frac{x}{8} + \dots,$$

due to Euler.

The discovery of logarithms

One great invention made early in the seventeenth century must be specially referred to; that of logarithms by John Napier (1550—1617). The special importance of this invention in relation to our subject is due to the fact of that essential connection between the numbers π and e which, after its discovery in the eighteenth century, dominated the later theory of the number π. The first announcement of the discovery was made in Napier's *Mirifici logarithmorum canonis descriptio* (Edinburgh, 1614), which contains an account of the nature of logarithms, and a table giving natural sines and their logarithms for every minute of the quadrant to seven or eight figures. These logarithms are not what are now called Napierian or natural logarithms (*i.e.* logarithms to the base e), although the former are closely related with the latter. The connection between the two is

$$L = 10^7 \log_e 10^7 - 10^7 . l, \text{ or } e^l = 10^7 e^{-\frac{L}{10^7}},$$

where l denotes the logarithm to the base e, and L denotes Napier's logarithm. It should be observed that in Napier's original theory of logarithms, their connection with the number e did not explicitly appear. The logarithm was not defined as the inverse of an exponential function; indeed the exponential function and even the exponential notation were not generally used by mathematicians till long afterwards.

Approximate constructions

A large number of approximate constructions for the rectification and quadrature of the circle have been given, some of which give very close approximations. It will suffice to give here a few examples of such constructions.

(1) The following construction for the approximate rectification of the circle was given by Kochansky (*Acta Eruditorum*, 1685).

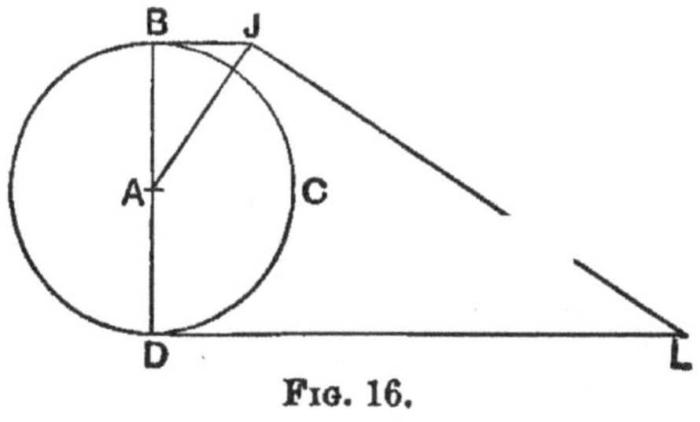

FIG. 16.

Let a length DL equal to 3 . radius be measured off on a tangent to the circle; let DAB be the diameter perpendicular to DL.

Let J be on the tangent at B, and such that $\angle BAJ = 30°$. Then JL is approximately equal to the semi-circular arc BCD. Taking the radius as unity, it can easily be proved that

$$JL = \sqrt{\frac{40}{3} - \sqrt{12}} = 3{\cdot}141533 \ldots$$

the correct value to four places of decimals.

(2) The value $\frac{355}{113} = 3{\cdot}141592 \ldots$ is correct to six decimal places. Since $\frac{355}{113} = 3 + \frac{4^2}{7^2 + 8^2}$, it can easily be constructed.

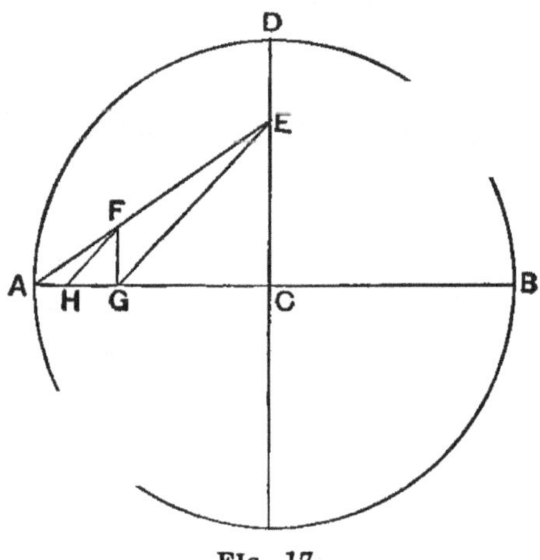

FIG. 17.

Let $CD = 1$, $CE = \frac{7}{8}$, $AF = \frac{1}{2}$; and let FG be parallel to CD and FH to EG; then $AH = \frac{4^2}{7^2 + 8^2}$.

This construction was given by Jakob de Gelder (*Grünert's Archiv*, vol. 7, 1849).

(3) At A make $AB = (2 + \frac{1}{5})$ radius on the tangent at A and let

$$BC = \tfrac{1}{5} \text{. radius.}$$

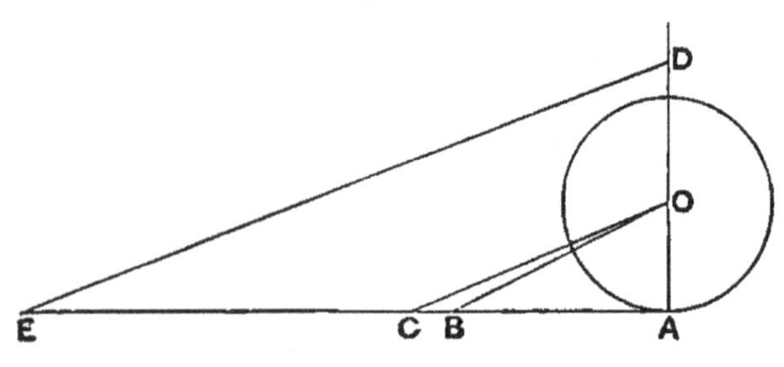

FIG. 18.

On the diameter through A take $AD = OB$, and draw DE parallel to OC. Then

$$\frac{AE}{AD} = \frac{AC}{AO} = \frac{13}{5}; \text{ therefore } AE = r \cdot \frac{13}{5}\sqrt{1 + \left(\frac{11}{5}\right)^2} = r \cdot \frac{13}{25}\sqrt{146};$$

thus $AE = r \cdot 6{\cdot}2831839\ldots$, so that AE is less than the circumference of the circle by less than two millionths of the radius.

The rectangle with sides equal to AE and half the radius r has very approximately its area equal to that of the circle. This construction was given by Specht (*Crelle's Journal*, vol. 3, p. 83).

(4) Let AOB be the diameter of a given circle. Let

$$OD = \tfrac{3}{5}r, \quad OF = \tfrac{3}{2}r, \quad OE = \tfrac{1}{2}r.$$

Describe the semi-circles DGE, AHF with DE and AF as diameters; and let the perpendicular to AB through O cut them in G and H respectively. The square of which the side is GH is approximately of area equal to that of the circle.

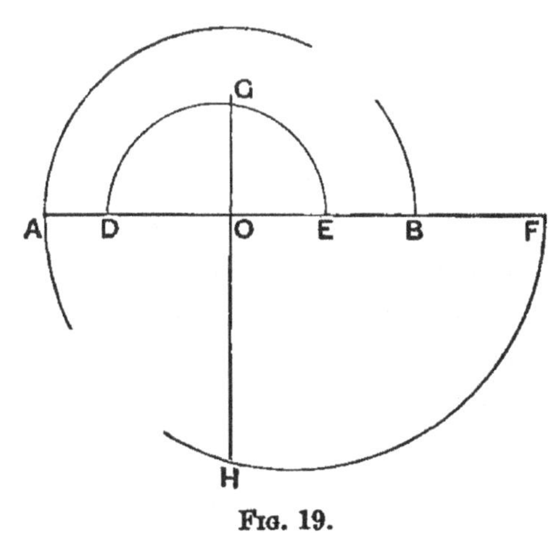

Fig. 19.

We find that $GH = r \cdot 1{\cdot}77246\ldots$, and since $\sqrt{\pi} = 1{\cdot}77245$ we see that GH is greater than the side of the square whose area is equal to that of the circle by less than two hundred thousandths of the radius.

CHAPTER III

THE SECOND PERIOD

The new Analysis

THE foundations of the new Analysis were laid in the second half of the seventeenth century when Newton (1642—1727) and Leibnitz (1646—1716) founded the Differential and Integral Calculus, the ground having been to some extent prepared by the labours of Huyghens, Fermat, Wallis, and others. By this great invention of Newton and Leibnitz, and with the help of the brothers James Bernouilli (1654—1705) and John Bernouilli (1667—1748), the ideas and methods of Mathematicians underwent a radical transformation which naturally had a profound effect upon our problem. The first effect of the new Analysis was to replace the old geometrical or semi-geometrical methods of calculating π by others in which analytical expressions formed according to definite laws were used, and which could be employed for the calculation of π to any assigned degree of approximation.

The work of John Wallis

The first result of this kind was due to John Wallis (1616—1703), Undergraduate at Emmanuel College, Fellow of Queens' College, and afterwards Savilian Professor of Geometry at Oxford. He was the first to formulate the modern arithmetic theory of limits, the fundamental importance of which, however, has only during the last half century received its due recognition; it is now regarded as lying at the very foundation of Analysis. Wallis gave in his *Arithmetica Infinitorum* the expression

$$\frac{\pi}{2}=\frac{2}{1}\cdot\frac{2}{3}\cdot\frac{4}{3}\cdot\frac{4}{5}\cdot\frac{6}{5}\cdot\frac{6}{7}\cdot\frac{8}{7}\cdot\frac{8}{9}\cdots$$

for π as an infinite product, and he shewed that the approximation obtained by stopping at any fraction in the expression on the right is in defect or in excess of the value $\frac{\pi}{2}$ according as the fraction is proper or improper. This expression was obtained by an ingenious method depending upon the expression for $\frac{\pi}{8}$ the area of a semi-circle of diameter 1 as the definite integral $\int_0^1 \sqrt{x-x^2}\,dx$. The expression has the advantage over that of Vieta that the operations required by it are all rational ones.

Lord Brouncker (1620—1684), the first President of the Royal Society, communicated without proof to Wallis the expression

$$\frac{4}{\pi} = 1 + \frac{1}{2+}\,\frac{9}{2+}\,\frac{25}{2+}\,\frac{49}{2+}\cdots,$$

a proof of which was given by Wallis in his *Arithmetica Infinitorum.* It was afterwards shewn by Euler that Wallis' formula could be obtained from the development of the sine and cosine in infinite products, and that Brouncker's expression is a particular case of much more general theorems.

The calculation of π by series

The expression from which most of the practical methods of calculating π have been obtained is the series which, as we now write it, is given by

$$\tan^{-1}x = x - \tfrac{1}{3}x^3 + \tfrac{1}{5}x^5 - \dots \qquad (-1 \leqq x \leqq 1).$$

This series was discovered by Gregory (1670) and afterwards independently by Leibnitz (1673). In Gregory's time the series was written as

$$a = t - \frac{t^3}{3r^2} + \frac{t^5}{5r^4} - \dots,$$

where a, t, r denote the length of an arc, the length of a tangent at one extremity of the arc, and the radius of the circle; the definition of the tangent as a ratio had not yet been introduced.

The particular case

$$\frac{\pi}{4} = 1 - \frac{1}{3} + \frac{1}{5} - \dots$$

is known as Leibnitz's series; he discovered it in 1674 and published it in 1682, with investigations relating to the representation of π, in his work "De vera proportione circuli ad quadratum circumscriptum in numeris rationalibus." The series was, however, known previously to Newton and Gregory.

By substituting the values $\frac{\pi}{6}, \frac{\pi}{8}, \frac{\pi}{10}, \frac{\pi}{12}$ in Gregory's series, the calculation of π up to 72 places was carried out by Abraham Sharp under instructions from Halley (Sherwin's ***Mathematical Tables,*** 1705, 1706).

The more quickly convergent series

$$\sin^{-1}x = x + \frac{1}{2}\frac{x^3}{3} + \frac{1.3}{2.4}\frac{x^5}{5} + \dots,$$

discovered by Newton, is troublesome for purposes of calculation, owing to the form of the coefficients. By taking $x = \frac{1}{2}$, Newton himself calculated π to 14 places of decimals.

Euler and others occupied themselves in deducing from Gregory's series formulae by which π could be calculated by means of rapidly converging series.

Euler, in 1737, employed special cases of the formula

$$\tan^{-1}\frac{1}{p} = \tan^{-1}\frac{1}{p+q} + \tan^{-1}\frac{q}{p^2+pq+1},$$

and gave the general expression

$$\tan^{-1}\frac{x}{y} = \tan^{-1}\frac{ax-y}{ay+x} + \tan^{-1}\frac{b-a}{ab+1} + \tan^{-1}\frac{c-b}{cb+1} + \dots,$$

from which more such formulae could be obtained. As an example, we have, if a, b, c, ... are taken to be the uneven numbers, and $\frac{x}{y} = 1$,

$$\frac{\pi}{4} = \tan^{-1}\frac{1}{2} + \tan^{-1}\frac{1}{2.4} + \tan^{-1}\frac{1}{2.9} + \dots.$$

In the year 1706, Machin (1680—1752), Professor of Astronomy in London, employed the series

$$\frac{\pi}{4} = 4\left(\frac{1}{5} - \frac{1}{3.5^3} + \frac{1}{5.5^5} - \frac{1}{7.5^7} + \dots\right)$$
$$- \left(\frac{1}{239} - \frac{1}{3.239^3} + \frac{1}{5.239^5} - \frac{1}{7.239^7} + \dots\right),$$

which follows from the relation

$$\frac{\pi}{4} = 4\tan^{-1}\frac{1}{5} - \tan^{-1}\frac{1}{239},$$

to calculate π to 100 places of decimals. This is a very convenient expression, because in the first series $\frac{1}{5}$, $\frac{1}{5^3}$, ... can be replaced by $\frac{4}{100}$, $\frac{64}{1000000}$, &c., and the second series is very rapidly convergent.

In 1719, de Lagny (1660—1734), of Paris, determined in two different ways the value of π up to 127 decimal places. Vega (1754—1802) calculated π to 140 places, by means of the formulae

$$\frac{\pi}{4} = 5\tan^{-1}\frac{1}{7} + 2\tan^{-1}\frac{3}{79} = 2\tan^{-1}\frac{1}{3} + \tan^{-1}\frac{1}{7},$$

due to Euler, and shewed that de Lagny's determination was correct with the exception of the 113th place, which should be 8 instead of 7.

Clausen calculated in 1847, 248 places of decimals by the use of Machin's formula and the formula

$$\frac{\pi}{4} = 2\tan^{-1}\frac{1}{3} + \tan^{-1}\frac{1}{7}.$$

In 1841, 208 places, of which 152 are correct, were calculated by Rutherford by means of the formula

$$\frac{\pi}{4} = 4\tan^{-1}\frac{1}{5} - \tan^{-1}\frac{1}{70} + \tan^{-1}\frac{1}{99}.$$

In 1844 an expert reckoner, Zacharias Dase, employed the formula

$$\frac{\pi}{4} = \tan^{-1}\frac{1}{2} + \tan^{-1}\frac{1}{5} + \tan^{-1}\frac{1}{8},$$

supplied to him by Prof. Schultz, of Vienna, to calculate π to 200 places of decimals, a feat which he performed in two months.

In 1853 Rutherford gave 440 places of decimals, and in the same year W. Shanks gave first 530 and then 607 places (*Proc. R. S.*, 1853).

Richter, working independently, gave in 1853 and 1855, first 333, then 400 and finally 500 places.

Finally, W. Shanks, working with Machin's formula, gave (1873—74) 707 places of decimals.

Another series which has also been employed for the calculation of π is the series

$$\tan^{-1} t = \frac{t}{1+t^2}\left\{1 + \frac{2}{3}\frac{t^2}{1+t^2} + \frac{2.4}{3.5}\left(\frac{t^2}{1+t^2}\right)^2 + \frac{2.4.6}{3.5.7}\left(\frac{t^2}{1+t^2}\right)^3 + \ldots\right\}.$$

This was given in the year 1755 by Euler, who, applying it in the formula

$$\pi = 20\tan^{-1}\tfrac{1}{7} + 8\tan^{-1}\tfrac{3}{79},$$

calculated π to 20 places, in one hour as he states. The same series was also discovered independently by Ch. Hutton (*Phil. Trans.*, 1776). It was later rediscovered by J. Thomson and by De Morgan.

An expression for π given by Euler may here be noticed; taking the identity

$$\tan^{-1}\frac{x}{2-x} = 2\int_0^x \frac{dx}{4+x^4} + 2\int_0^x \frac{x\,dx}{4+x^4} + \int_0^x \frac{x^2 dx}{4+x^4},$$

he developed the integrals in series, then put $x = \frac{1}{2}$, $x = \frac{1}{4}$, obtaining series for $\tan^{-1}\frac{1}{3}$, $\tan^{-1}\frac{1}{7}$, which he substituted in the formula

$$\frac{\pi}{4} = 2\tan^{-1}\frac{1}{3} + \tan^{-1}\frac{1}{7}.$$

In China a work was published by Imperial order in 1713 which contained a chapter on the quadrature of the circle where the first 19 figures in the value of π are given.

At the beginning of the eighteenth century, analytical methods were introduced into China by Tu Tê-mei (Pierre Jartoux) a French missionary; it is, however, not known how much of his work is original, or whether he borrowed the formulae he gave directly from European Mathematicians.

One of his series

$$\pi = 3\left(1 + \frac{1^2}{4.6} + \frac{1^2.3^2}{4.6.8.10} + \frac{1^2.3^2.5^2}{4.6.8.10.12.14} + \dots\right)$$

was employed at the beginning of the nineteenth century by Chu-Hung for the calculation of π. By this means 25 correct figures were obtained.

Tsêng Chi-hung, who died in 1877, published values of π and $1/\pi$ to 100 places. He is said to have obtained his value of π in a month, by means of the formula

$$\frac{\pi}{4} = \tan^{-1}\frac{1}{2} + \tan^{-1}\frac{1}{3}$$

and Gregory's series.

In Japan, where a considerable school of Mathematics was developed in the eighteenth century, π was calculated by Takebe in 1722 to 41 places, by employment of the regular 1024agon. It was calculated by Matsunaga in 1739 to 50 places by means of the same series as had been employed by Chu-Hung.

The rational values $\pi = \frac{5419351}{1725033}$, $\pi = \frac{428224593349304}{136308121570117}$, correct to 12 and 30 decimal places respectively, were given by Arima in 1766.

Kurushima Yoshita (died 1757) gave for π^2 the approximate values $\frac{227}{23}$, $\frac{10748}{1089}$, $\frac{10975}{1112}$, $\frac{98548}{9985}$.

Tanyem Shŏkei published in 1728 the series

$$\pi^2 = 8\left(1 + \frac{1}{6} + \frac{1.4}{6.15} + \frac{1.4.9}{6.15.28} + \dots\right),$$

$$\pi^2 = 4\left(1 + \frac{2}{6} + \frac{2.8}{6.15} + \frac{2.8.18}{6.15.28} + \dots\right)$$

due to Takebe, and ultimately to Jartoux.

The following series published in 1739 by Matsunaga may be mentioned:

$$\pi^2 = 9\left(1 + \frac{1^2}{3.4} + \frac{1^2.2^2}{3.4.5.6} + \frac{1^2.2^2.3^2}{3.4.5.6.7.8} + \dots\right).$$

The work of Euler

Developments of the most far-reaching importance in connection with our subject were made by Leonhard Euler, one of the greatest Analysts of all time, who was born at Basel in 1707 and died at St Petersburg in 1783. With his vast influence on the development of Mathematical Analysis in general it is impossible here to deal, but some account must be given of those of his discoveries which come into relation with our problem.

The very form of modern Trigonometry is due to Euler. He introduced the practice of denoting each of the sides and angles of a triangle by a single letter, and he introduced the short designation of the trigonometrical ratios by sin α, cos α, tan α, &c. Before Euler's time there was great prolixity in the statement of propositions, owing to the custom of denoting these expressions by words, or by letters specially introduced in the statement. The habit of denoting the ratio of the circumference to the diameter of a circle by the letter π, and the base of the natural system of logarithms by e, is due to the influence of the works of Euler, although the notation π appears as early as 1706, when it was used by William Jones in the *Synopsis palmariorum Matheseos*. In Euler's earlier work he frequently used p instead of π, but by about 1740 the letter π was used not only by Euler but by other Mathematicians with whom he was in correspondence.

A most important improvement which had a great effect not only upon the form of Trigonometry but also on Analysis in general was the introduction by Euler of the definition of the trigonometrical ratios in order to replace the old sine, cosine, tangent, &c., which were the lengths of straight lines connected with the circular arc. Thus these trigonometrical ratios became functions of an angular magnitude, and therefore numbers, instead of lengths of lines related by equations with the radius of the circle. This very important improvement was not generally introduced into our text books until the latter half of the nineteenth century.

This mode of regarding the trigonometrical ratios as analytic functions led Euler to one of his greatest discoveries, the connection of these functions with the exponential function. On the basis of the definition of e^z by means of the series

$$1 + z + \frac{z^2}{2\,!} + \frac{z^3}{3\,!} + \dots,$$

he set up the relations

$$\cos x = \frac{e^{ix} + e^{-ix}}{2}, \qquad \sin x = \frac{e^{ix} - e^{-ix}}{2},$$

which can also be written

$$e^{ix} = \cos x + i \sin x, \quad e^{-ix} = \cos x - i \sin x.$$

The relation $e^{i\pi} = -1$, which Euler obtained by putting $x = \pi$, is the fundamental relation between the two numbers π and e which was indispensable later on in making out the true nature of the number π.

In his very numerous memoirs, and especially in his great work, *Introductio in analysin infinitorum* (1748), Euler displayed the most wonderful skill in obtaining a rich harvest of results of great interest, largely dependent on his theory of the exponential function. Hardly any other work in the history of Mathematical Science gives to the reader so strong an impression of the genius of the author as the *Introductio.* Many of the results given in that work are obtained by bold generalizations, in default of proofs which would now be regarded as completely rigorous; but this it has in common with a large part of all Mathematical discoveries, which are often due to a species of divining intuition, the rigorous demonstrations and the necessary restrictions coming later. In particular there may be mentioned the expressions for the sine and cosine functions as infinite products, and a great number of series and products deduced from these expressions; also a number of expressions relating the number e with continued fractions which were afterwards used in connection with the investigation of the nature of that number.

Great as the progress thus made was, regarded as preparatory to a solution of our problem, nothing definite as to the true nature of the number π was as yet established, although Mathematicians were convinced that e and π are not roots of algebraic equations. Euler himself gave expression to the conviction that this is the case. Somewhat later, Legendre gave even more distinct expression to this view in his *Éléments de Géométrie* (1794), where he writes: "It is probable that the number π is not even contained among the algebraical irrationalities, *i.e.* that it cannot be a root of an algebraical equation with a finite number of terms, whose coefficients are rational. But it seems to be very difficult to prove this strictly."

CHAPTER IV

THE THIRD PERIOD

The irrationality of π and e

THE third and final period in the history of the problem is concerned with the investigation of the real nature of the number π. Owing to the close connection of this number with the number e, the base of natural logarithms, the investigation of the nature of the two numbers was to a large extent carried out at the same time.

The first investigation, of fundamental importance, was that of J. H. Lambert (1728—1777), who in his "Mémoire sur quelques propriétés remarquables des quantités transcendentes circulaires et logarithmiques" (*Hist. de l'Acad. de Berlin*, 1761, printed in 1768), proved that e and π are irrational numbers. His investigations are given also in his treatise *Vorläufige Kenntnisse für die, so die Quadratur und Rektification des Zirkels suchen*, published in 1766.

He obtained the two continued fractions

$$\frac{e^x - 1}{e^x + 1} = \frac{1}{2/x +} \frac{1}{6/x +} \frac{1}{10/x +} \frac{1}{14/x + \ldots},$$

$$\tan x = \frac{1}{1/x -} \frac{1}{3/x -} \frac{1}{5/x -} \frac{1}{7/x - \ldots},$$

which are closely related with continued fractions obtained by Euler, but the convergence of which Euler had not established. As the result of an investigation of the properties of these continued fractions, Lambert established the following theorems:

(1) If x is a rational number, different from zero, e^x cannot be a rational number.

(2) If x is a rational number, different from zero, $\tan x$ cannot be a rational number.

If $x = \frac{1}{4}\pi$, we have $\tan x = 1$, and therefore $\frac{1}{4}\pi$, or π, cannot be a rational number.

It has frequently been stated that the first rigorous proof of Lambert's results is due to Legendre (1752—1833), who proved these theorems in his *Éléments de Géométrie* (1794), by the same method, and added a proof that π^2 is an irrational number. The essential rigour of Lambert's proof has however been pointed out by Pringsheim (*Münch. Akad. Ber.*, Kl. 28, 1898), who has supplemented the investigation in respect of the convergence.

A proof of the irrationality of π and π^2 due to Hermite (*Crelle's Journal*, vol. 76, 1873) is of interest, both in relation to the proof of Lambert, and as containing the germ of the later proof of the transcendency of e and π.

A simple proof of the irrationality of e was given by Fourier (Stainville, *Mélanges d'analyse*, 1815), by means of the series

$$1+\frac{1}{1!}+\frac{1}{2!}+\frac{1}{3!}+\dots$$

which represents the number. This proof can be extended to shew that e^2 is also irrational. On the same lines it was proved by Liouville (1809—1882) (*Liouville's Journal*, vol. 5, 1840) that neither e nor e^2 can be a root of a quadratic equation with rational coefficients. This last theorem is of importance as forming the first step in the proof that e and π cannot be roots of any algebraic equation with rational coefficients. The probability had been already recognized by Legendre that there exist numbers which have this property.

Existence of transcendental numbers

The confirmation of this surmised existence of such numbers was obtained by Liouville in 1840, who by an investigation of the properties of the convergents of a continued fraction which represents a root of an algebraical equation, and also by another method, proved that numbers can be defined which cannot be the root of any algebraical equation with rational coefficients.

The simpler of Liouville's methods of proving the existence of such numbers will be here given.

Let x be a real root of the algebraic equation

$$ax^n+bx^{n-1}+cx^{n-2}+\dots=0,$$

with coefficients which are all positive or negative integers. We shall assume that this equation has all its roots unequal; if it had equal roots we might suppose it to be cleared of them in the usual manner.

Let the other roots be denoted by $x_1, x_2, \dots x_{n-1}$; these may be real or complex. If $\frac{p}{q}$ be any rational fraction, we have

$$\frac{p}{q} - x = \frac{ap^n + bp^{n-1}q + cp^{n-2}q^2 + \dots}{q^n \cdot a\left(\frac{p}{q} - x_1\right)\left(\frac{p}{q} - x_2\right)\dots\left(\frac{p}{q} - x_n\right)}.$$

If now we have a sequence of rational fractions converging to the value x as limit, but none of them equal to x, and if $\frac{p}{q}$ be one of these fractions,

$$\left(\frac{p}{q} - x_1\right)\left(\frac{p}{q} - x_2\right)\dots\left(\frac{p}{q} - x_n\right)$$

approximates to the fixed number

$$(x - x_1)(x - x_2)\dots(x - x_n).$$

We may therefore suppose that for all the fractions $\frac{p}{q}$,

$$a\left(\frac{p}{q} - x_1\right)\left(\frac{p}{q} - x_2\right)\dots\left(\frac{p}{q} - x_n\right)$$

is numerically less than some fixed positive number A. Also

$$ap^n + bp^{n-1}q + \dots$$

is an integer numerically $\geqq 1$; therefore

$$\left|\frac{p}{q} - x\right| > \frac{1}{Aq^n}.$$

This must hold for all the fractions $\frac{p}{q}$ of such a sequence, from and after some fixed element of the sequence, for some fixed number A. If now a number x can be so defined such that, however far we go in the sequence of fractions $\frac{p}{q}$, and however A be chosen, there exist fractions belonging to the sequence for which $\left|\frac{p}{q} - x\right| < \frac{1}{Aq^n}$, it may be concluded that x cannot be a root of an equation of degree n with integral coefficients. Moreover, if we can shew that this is the case whatever value n may have, we conclude that x cannot be a root of any algebraic equation with rational coefficients.

Consider a number

$$x = \frac{k_1}{r^{1!}} + \frac{k_2}{r^{2!}} + \dots + \frac{k_m}{r^{m!}} + \dots,$$

where the integers $k_1, k_2, \ldots k_m, \ldots$ are all less than the integer r, and do not all vanish from and after a fixed value of m.

Let
$$\frac{p}{q} = \frac{k_1}{r^{1!}} + \frac{k_2}{r^{2!}} + \ldots + \frac{k_m}{r^{m!}},$$

then $\frac{p}{q}$ continually approaches x as m is increased. We have

$$x - \frac{p}{q} = \frac{k_{m+1}}{r^{(m+1)!}} + \frac{k_{m+2}}{r^{(m+2)!}} + \ldots$$
$$< r\left(\frac{1}{r^{(m+1)!}} + \frac{1}{r^{(m+2)!}} + \ldots\right)$$
$$< \frac{2r}{q^{m+1}}, \text{ since } q = r^{m!}.$$

It is clear that, whatever values A and n may have, if m, and therefore q, is large enough, we have $\frac{2r}{q^{m+1}} < \frac{1}{Aq^n}$; and thus the relation $\left|\frac{p}{q} - x\right| > \frac{1}{Aq^n}$ is not satisfied for all the fractions $\frac{p}{q}$. The numbers x so defined are therefore transcendental. If we take $r = 10$, we see how to define transcendental numbers that are expressed as decimals.

This important result provided a complete justification of the division of numbers into two classes, algebraical numbers, and transcendental numbers; the latter being characterized by the property that such a number cannot be a root of an algebraical equation of any degree whatever, of which the coefficients are rational numbers.

A proof of this fundamentally important distinction, depending on entirely different principles, was given by G. Cantor (*Crelle's Journal*, vol. 77, 1874) who shewed that the algebraical numbers form an enumerable aggregate, that is to say that they are capable of being counted by means of the integer sequence 1, 2, 3, ..., whereas the aggregate of all real numbers is not enumerable. He shewed how numbers can be defined which certainly do not belong to the sequence of algebraic numbers, and are therefore transcendental.

This distinction between algebraic and transcendental numbers being recognized, the question now arose, as regards any particular number defined in an analytical manner, to which of the two classes it belongs; in particular whether π and e are algebraic or transcendental. The difficulty of answering such a question arises from the fact that the recognition of the distinction between the two classes of numbers does not of itself provide a readily applicable criterion by the use of which the question may be answered in respect of a particular number.

The scope of Euclidean determinations

Before proceeding to describe the manner in which it was finally shewn that the number π is a transcendental number, it is desirable to explain in what way this result is connected with the problems of the quadrature and rectification of the circle by means of Euclidean determinations.

The development of Analytical Geometry has made it possible to replace every geometrical problem by a corresponding analytical one which involves only numbers and their relations. As we have already remarked, every Euclidean problem of what is called construction consists essentially in the determination of one or more points which shall satisfy certain prescribed relations with regard to a certain finite number of assigned points, the data of the problem. Such a problem has as its analytical counterpart the determination of a number, or a finite set of numbers, which shall satisfy certain prescribed relations relatively to a given set of numbers. The determination of the required numbers is always made by means of a set of algebraical equations.

The development of the theory of algebraical equations, especially that due to Abel, Gauss, and Galois, led the Mathematicians of the last century to scrutinize with care the limits of the possibility of solving geometrical problems subject to prescribed limitations as to the nature of the geometrical operations regarded as admissible. In particular, it has been ascertained what classes of geometrical problems are capable of solution when operations equivalent in practical geometry to the use of certain instruments are admitted*. The investigations have led to the discovery of cases such as that of inscribing a regular polygon of 17 sides in a circle, in which a problem, not previously known to be capable of solution by Euclidean means, has been shewn to be so.

We shall here give an account of as much of the theory of this subject as is necessary for the purpose of application to the theory of the quadrature and rectification of the circle.

In the first place we observe that, having given two or more points in a plane, a Cartesian set of axes can be constructed by means of a

* An interesting detailed account of investigations of this kind will be found in Enriques' *Questions of Elementary Geometry*, German Edition, 1907.

Euclidean construction, for example by bisecting the segment of the line on which two of the given points are incident, and then determining a perpendicular to that segment. We may therefore assume that a given set of points, the data of a Euclidean problem, are specified by means of a set of numbers, the coordinates of these points.

The determination of a required point P is, in a Euclidean problem, made by means of a finite number of applications of the three processes, (1) of determining a new point as the intersection of straight lines given each by a pair of points already determined, (2) of determining a new point as an intersection of a straight line given by two points and a circle given by its centre and one point on the circumference, all four points having been already determined, and (3) of determining a new point as an intersection of two circles which are determined by four points already determined.

In the analytical interpretation we have an original set of numbers $a_1, a_2, \dots a_{2r}$ given, the coordinates of the r given points; $(r \geqq 2)$. At each successive stage of the geometrical process we determine two new numbers, the coordinates of a fresh point.

When a certain stage of the process has been completed, the data for the next step consist of numbers $(a_1, a_2, \dots a_{2n})$ containing the original data and those numbers which have been already ascertained by the successive stages of the process already carried out.

If (1) is employed for the next step of the geometrical process, the new point determined by that step corresponds to numbers determined by two equations

$$Ax + By + C = 0, \quad A'x + B'y + C' = 0,$$

where A, B, C, A', B', C' are rational functions of eight of the numbers $(a_1, a_2, \dots a_{2n})$. Therefore x, y the coordinates of the new point determined by this step are rational functions of $a_1, a_2, \dots a_{2n}$.

In order to get the data for the next step afterwards, we have only to add to $a_1, a_2, \dots a_{2n}$ these two rational functions of eight of them.

If case (2) is employed, the next point is determined by two equations of the form

$$(x - a_p)^2 + (y - a_q)^2 = (a_r - a_s)^2 + (a_t - a_u)^2,$$

$$y = mx + n,$$

where m, n are rational functions of four of the numbers $a_1, a_2, \dots a_{2n}$. On elimination of y, we have a quadratic equation for x; and thus x is determined as a quadratic irrational function of $(a_1, a_2, \dots a_{2n})$, of the form $A \pm \sqrt{B}$, where A and B are rational functions; it is clear that y will be determined in a similar way.

If, in the new step, (3) is employed, the equations for determining (x, y) consist of two equations of the form

$$(x-a_p)^2+(y-a_q)^2=(a_r-a_s)^2+(a_t-a_u)^2;$$

on subtracting these equations, we obtain a linear equation, and thus it is clear that this case is essentially similar to that in which (2) is employed, so far as the form of x, y is concerned.

Since the determination of a required point P is to be made by a finite number of such steps, we see that the coordinates of P are determined by means of a finite succession of operations on

$$(a_1, a_2, \dots a_{2r}),$$

the coordinates of the points; each of these operations consists either of a rational operation, or of one involving the process of taking a square root of a rational function as well as a rational operation.

We have now established the following result:

In order that a point P can be determined by the Euclidean mode it is necessary and sufficient that its coordinates can be expressed as such functions of the coordinates $(a_1, a_2, \dots a_{2r})$ of the given points of the problem, as involve the successive performance, a finite number of times, of operations which are either rational or involve taking a square root of a rational function of the elements already determined.

That the condition stated in this theorem is necessary has been proved above; that it is sufficient is seen from the fact that a single rational operation, and the single operation of taking a square root of a number already known, are both operations which correspond to possible Euclidean determinations.

The condition stated in the result just obtained may be put in another form more immediately available for application. The expression for a coordinate x of the point P may, by the ordinary processes for the simplification of surd expressions, by getting rid of surds from the denominators of fractions, be reduced to the form

$$x=a+b\sqrt{c_1 \pm \sqrt{c_2 \pm \sqrt{c_3+\dots}}}+b'\sqrt{c_1' \pm \sqrt{c_2' \pm \sqrt{c_3'+\dots}}}+\dots,$$

where all the numbers

$$a, b, c_1, c_2, \dots, b', c_1', c_2', \dots$$

are rational functions of the given numbers $(a_1, a_2, \dots a_{2r})$, and the number of successive square roots is in every term finite. Let m be the greatest number of successive square roots in any term of x; this may be called the rank of x. We may then write

$$x = a + b\sqrt{B} + b'\sqrt{B'} + \ldots,$$

where B, B', ... are all of rank not greater than $m-1$. We can form an equation which x satisfies, and such that all its coefficients are rational functions of a, b, b', B, B' ...; for $\sqrt{B}$ may be eliminated by taking $(x - a - b'\sqrt{B'} - \ldots)^2 = b^2B$, and this is of the form

$$P_2 + \sqrt{B'}P_2' = 0,$$

from which we form the biquadratic

$$P_2^2 - B'P_2'^2 = 0,$$

in which $\sqrt{B'}$ does not occur. Proceeding in this way we obtain an equation in x of degree some power of 2, and of which the coefficients are rational functions of a, b, B, B', ..., and are therefore of rank $\leqq m-1$. This equation is of the form

$$L_1 x^{2^s} + L_2 x^{2^s - 1} + \ldots = 0,$$

where L_1, L_2, ... are at most of rank $m-1$. If L_1, L_2, ... involve a radical $\sqrt{K}$, the equation is of the form

$$\sqrt{K}(b_1 x^{2^s} + \ldots) + (b_1' x^{2^s} + \ldots) = 0,$$

and we can as before reduce this to an equation of degree 2^{s+1} in which $\sqrt{K}$ does not occur; by repeating the process for each radical like $\sqrt{K}$, we may eliminate them all, and finally obtain an equation such that the rank of every coefficient is $\leqq m-2$. By continual repetition of this procedure we ultimately reach an equation, such that the coefficients are all of rank zero, *i.e.* rational functions of $(a_1, a_2, \ldots a_{2r})$. We now see that the following result has been established:

In order that a point P may be determinable by Euclidean procedure it is necessary that each of its coordinates be a root of an equation of some degree, a power of 2, of which the coefficients are rational functions of $(a_1, a_2, \ldots a_{2r})$, *the coordinates of the points given in the data of the problem.*

From our investigation it is clear that only those algebraic equations which are obtainable by elimination from a sequence of linear and quadratic equations correspond to possible Euclidean problems.

The quadratic equations must consist of sets, those in the first set having coefficients which are rational functions of the given numbers, those in the second set having coefficients of rank at most 1; in the next set the coefficients have rank at most 2, and so on.

The criterion thus obtained is sufficient, whenever it can be applied, to determine whether a proposed Euclidean problem is a possible one or not.

In the case of the rectification of the circle, we may assume that the data of the problem consist simply of the two points (0, 0) and (1, 0), and that the point to be determined has the coordinates $(\pi, 0)$. This will, in accordance with the criterion obtained, be a possible problem only if π is a root of an algebraic equation with rational coefficients, of that special class which has roots expressible by means of rational numbers and numbers obtainable by successive operations of taking the square roots. The investigations of Abel have shewn that this is only a special class of algebraic equations.

As we shall see, it is now known that π, being transcendental, is not a root of any algebraic equation at all, and therefore in accordance with the criterion is not determinable by Euclidean construction. The problems of duplication of the cube, and of the trisection of an angle, although they lead to algebraic equations, are not soluble by Euclidean constructions, because the equations to which they lead are not in general of the class referred to in the above criterion.

The transcendence of π

In 1873 Ch. Hermite* succeeded in proving that the number e is transcendental, that is that no equation of the form

$$ae^m + be^n + ce^r + \ldots = 0$$

can subsist, where $m, n, r, \ldots a, b, c, \ldots$ are whole numbers. In 1882, the more general theorem was stated by Lindemann that such an equation cannot hold, when $m, n, r, \ldots a, b, c, \ldots$ are algebraic numbers, not necessarily real; and the particular case that $e^{ix} + 1 = 0$ cannot be satisfied by an algebraic number x, and therefore that π is not algebraic, was completely proved by Lindemann†.

Lindemann's general theorem may be stated in the following precise form:

If $x_1, x_2, \ldots x_n$ are any real or complex algebraical numbers, all distinct, and $p_1, p_2, \ldots p_n$ are n algebraical numbers at least one of which is different from zero, then the sum

$$p_1 e^{x_1} + p_2 e^{x_2} + \ldots + p_n e^{x_n}$$

is certainly different from zero.

* "Sur la fonction exponentielle," *Comptes Rendus*, vol. 77, 1873.

† *Ber. Akad. Berlin*, 1882.

The particular case of this theorem in which

$$n=2, \quad x_1=ix, \quad x_2=0, \quad p_1=p_2=1,$$

shews that $e^{ix}+1$ cannot be zero if x is an algebraic number, and thus that, since $e^{i\pi}+1=0$, it follows that *the number π is transcendental.*

From the general theorem there follow also the following important results:

(1) Let $n=2$, $p_1=1$, $p_2=-a$, $x_1=x$, $x_2=0$; then the equation $e^x-a=0$ cannot hold if x and a are both algebraic numbers and x is different from zero. Hence *the exponential e^x is transcendent if x is an algebraic number different from zero. In particular e is transcendent.* Further, *the natural logarithm of an algebraic number different from zero is a transcendental number. The transcendence of $i\pi$ and therefore of π is a particular case of this theorem.*

(2) Let $n=3$, $p_1=-i$, $p_2=i$, $p_3=-2a$, $x_1=ix$, $x_2=-ix$, $x_3=0$; it then follows that the equation $\sin x=a$ cannot be satisfied if a and x are both algebraic numbers different from zero. Hence, *if $\sin x$ is algebraic, x cannot be algebraic, unless $x=0$, and if a is algebraic, $\sin^{-1} a$ cannot be algebraic, unless $a=0$.*

It is easily seen that a similar theorem holds for the cosine and the other trigonometrical functions.

The fact that π is a transcendental number, combined with what has been established above as regards the possibility of Euclidean constructions or determinations with given data, affords the final answer to the question whether the quadrature or the rectification of the circle can be carried out in the Euclidean manner.

The quadrature and the rectification of a circle whose diameter is given are impossible, as problems to be solved by the processes of Euclidean Geometry, in which straight lines and circles are alone employed in the constructions.

It appears, however, that the transcendence of π establishes the fact that *the quadrature or the rectification of a circle whose diameter is given are impossible by a construction in which the use only of algebraic curves is allowed.*

The special case (2) of Lindemann's theorem throws light on the interesting problems of the rectification of arcs of circles and of the quadrature of sectors of circles. If we take the radius of a circle to be unity then $2\sin\frac{1}{2}x$ is the length of the chord of an arc of which the length is x. It has been shewn that $2\sin\frac{1}{2}x$ and x cannot both be algebraic, unless $x=0$. We have therefore the following result:

If the chord of a circle bears to the diameter a ratio which is algebraic, then the corresponding arc is not rectifiable by any construction in which algebraic curves alone are employed; neither can the quadrature of the corresponding sector of the circle be carried out by such a construction.

The method employed by Hermite and Lindemann was of a complicated character, involving the use of complex integration. The method was very considerably simplified by Weierstrass*, who gave a complete proof of Lindemann's general theorem.

Proofs of the transcendence of e and π, progressively simple in character, were given by Stieltjes†, Hilbert, Hurwitz and Gordan‡, Mertens§, and Vahlen‖.

All these proofs consist of a demonstration that an equation which is linear in a number of exponential functions, such that the coefficients are whole numbers, and the exponents algebraic numbers, is impossible. By choosing a multiplier of the equation of such a character that its employment reduces the given equation to the equation of the sum of a non-vanishing integer and a number proved to lie numerically between 0 and 1 to zero, the impossibility is established.

Simplified presentations of the proofs will be found in Weber's *Algebra*, in Enriques' *Questions of Elementary Geometry* (German Edition, 1907), in Hobson's *Plane Trigonometry* (second edition, 1911), and in Art. IX. of the "Monographs on Modern Mathematics," edited by J. W. A. Young.

Proof of the transcendence of π

The proof of the transcendence of π which will here be given is founded upon that of Gordan.

(1) Let us assume that, if possible, π is a root of an algebraical equation with integral coefficients; then $i\pi$ is also a root of such an equation.

Assume that $i\pi$ is a root of the equation

$$C(x-\alpha_1)(x-\alpha_2)\ldots(x-\alpha_s)=0,$$

where all the coefficients

$$C,\ C\Sigma\alpha,\ C\Sigma\alpha_r\alpha_s,\ \ldots,\ C\alpha_1\alpha_2\ldots\alpha_s$$

* *Ber. Akad. Berlin*, 1885.

† *Comptes Rendus*, Paris Acad. 1890.

‡ These proofs are to be found in the *Math. Annalen*, vol. 43 (1896), by Hilbert, Hurwitz and Gordan.

§ *Wiener Ber.* Kl. cv. II*a* (1896).

‖ *Math. Annalen*, vol. 53 (1900).

are positive or negative integers (including zero); thus one of the numbers $\alpha_1, \ldots \alpha_s$ is $i\pi$.

From Euler's equation $e^{i\pi} + 1 = 0$, we see that the relation

$$(1 + e^{\alpha_1})(1 + e^{\alpha_2}) \ldots (1 + e^{\alpha_s}) = 0$$

must hold, since one of the factors vanishes. If we multiply out the factors in this equation, it clearly takes the form

$$A + e^{\beta_1} + e^{\beta_2} + \ldots + e^{\beta_n} = 0,$$

where A is some positive integer ($\geqq 1$), being made up of 1 together with those terms, if any, which are of the form $e^{\alpha_p + \alpha_q + \ldots}$, where

$$\alpha_p + \alpha_q + \ldots = 0.$$

(2) A symmetrical function consisting of the sum of the products taken in every possible way, of a fixed number of the numbers $C\alpha_1, C\alpha_2, \ldots C\alpha_s$, is an integer. It will be proved that the symmetrical functions of $C\beta_1, C\beta_2, \ldots C\beta_n$ have the same property. In order to prove this we have need of the following lemma:

A symmetrical function consisting of the sums of the products taken p together of $\alpha + \beta + \gamma + \ldots$ letters

$$x_1, x_2, \ldots x_\alpha; \quad y_1, y_2, \ldots y_\beta; \quad z_1, z_2, \ldots z_\gamma; \text{ \&c.},$$

belonging to any number of separate sets, can be expressed in terms of symmetrical functions of the letters in the separate sets.

It will be sufficient to prove this in the case in which there are only two sets of letters, the extension to the general case being then obvious.

Denote by $\sum\limits_p P(x, y)$ the sum of the products which we require to express; and denote by $\sum\limits_r P(x)$ the sum of the products of r dimensions of the letters $x_1, x_2, \ldots x_\alpha$ only. In case $p \leqq \alpha$, we see that

$$\sum_p P(x, y) = \sum_p P(x) + \sum_1 P(y) \sum_{p-1} P(x) + \sum_2 P(y) \sum_{p-2} P(x) + \ldots;$$

in case $p > \alpha$, we see that

$$\sum_p P(x, y)$$
$$= \sum_\alpha P(x) \sum_{p-\alpha} P(y) + \sum_{\alpha-1} P(x) \sum_{p-\alpha+1} P(y) + \sum_{\alpha-2} P(x) \sum_{p-\alpha+2} P(y) + \ldots;$$

and the terms on the right-hand side involve in each case only symmetrical functions of the letters of the two separate sets; thus the lemma is established.

To apply this lemma, we observe that the numbers β fall into separate sets, according to the way they are formed from the letters α.

The general value of β consists of the sum of r of the letters $\alpha_1, \alpha_2, \dots \alpha_s$; and we consider those values of β that correspond to a fixed value of r to belong to one set. It is clear that a symmetrical function of those letters β which belong to one and the same set is expressible as a symmetrical function of $\alpha_1, \alpha_2, \dots \alpha_s$; therefore a symmetrical function of the products $C\beta$, where all the β's belong to one and the same set, is in virtue of what has been established in (1) an integer. Applying the above lemma to all the n numbers $C\beta$, we see that the symmetrical products formed by all the numbers $C\beta$ are integral, or zero. We have supposed those of the numbers β which vanish to be suppressed and the corresponding exponentials to be absorbed in the integer A; whether this is done before or after the symmetrical functions of $C\beta$ are formed makes no difference, so that the above reasoning applies to the numbers $C\beta$ when those of them which vanish are removed.

(3) Let p be a prime number greater than all the numbers A, n, C $| C^n \beta_1 \beta_2 \dots \beta_n |$; and let

$$\phi(x) = \frac{x^{p-1}}{(p-1)!} C^{np+p-1} \{(x-\beta_1)(x-\beta_2)\dots(x-\beta_n)\}^p.$$

We observe that $\phi(x)$ is of the form

$$\frac{(Cx)^{p-1}}{(p-1)!} [(Cx)^n - q_1 (Cx)^{n-1} + q_2 (Cx)^{n-2} - \dots + (-1)^n q_n]^p,$$

where $q_1, q_2, \dots q_n$ are integers. The function $\phi(x)$ may be expressed in the form

$$\phi(x) = c_{p-1} x^{p-1} + c_p x^p + \dots + c_{np+p-1} x^{np+p-1},$$

where $c_{p-1}(p-1)!$, $c_p p!$, ... are integral.

We see that $\phi^{p-1}(0) = (-1)^{np} C^{p-1} q_n^p$, which is an integer not divisible by p.

Also $\phi^p(0)$ is the value when $x = 0$ of

$$pC^{p-1} \frac{d}{dx} [(Cx)^n - q_1 (Cx)^{n-1} + \dots]^p$$

and is clearly an integer divisible by p. We see also that

$$\phi^{(p+1)}(0), \quad \phi^{(p+2)}(0), \dots \phi^{np+p-1}(0)$$

are all multiples of p.

Further if $m \leqq n$, $\phi(\beta_m)$, $\phi'(\beta_m)$, ... $\phi^{(p-1)}(\beta_m)$ all vanish, and

$$\sum_{m=1}^{m=n} \phi^{(p)}(\beta_m), \quad \sum_{m=1}^{m=n} \phi^{(p+1)}(\beta_m), \dots \sum_{m=1}^{m=n} \phi^{(mp+p-1)}(\beta_m)$$

are all integers divisible by p. This follows from the fact that $\sum_{m=1}^{m=n}(C\beta_m)^r$ is expressible in terms of those symmetrical functions which consist of the sums of products of the numbers $C\beta_1$, $C\beta_2$, ...; and these expressions have integral values.

(4) Let K_p denote the integer

$$(p-1)!\,c_{p-1}+p!\,c_p+\ldots+(np+p-1)!,$$

which may be written in the form

$$\phi^{(p-1)}(0)+\phi^{(p)}(0)+\ldots+\phi^{(np+p-1)}(0).$$

In virtue of what has been established in (3) as to the values of

$$\phi^{(p-1)}(0),\quad \phi^{(p)}(0),\ \ldots$$

we see that K_pA is not a multiple of p.

We examine the form to which the equation

$$A+e^{\beta_1}+e^{\beta_2}+\ldots+e^{\beta_n}=0$$

is reduced by multiplying all the terms by K_p.

We have

$$K_pe^{\beta_m}=\sum_{r=p-1}^{r=np+p-1}c_r\left\{\beta_m{}^r+r\beta_m{}^{r-1}+r(r-1)\beta_m{}^{r-2}+\ldots+r!\right.$$
$$\left.+\frac{\beta_m{}^{r+1}}{r+1}+\frac{\beta_m{}^{r+2}}{(r+1)(r+2)}+\ldots\right\}$$
$$=\phi(\beta_m)+\phi'(\beta_m)+\ldots+\phi^{np+p-1}(\beta_m)$$
$$+\sum_{r=p-1}^{r=np+p-1}c_r\beta_m{}^r\left\{\frac{\beta_m}{r+1}+\frac{\beta_m{}^2}{(r+1)(r+2)}+\ldots\right\}.$$

The modulus of the sum of the series

$$\frac{\beta_m}{r+1}+\frac{\beta_{m+1}}{(r+1)(r+2)}+\ldots$$

does not exceed

$$\frac{|\beta_m|}{r+1}+\frac{|\beta_m|^2}{(r+1)(r+2)}+\ldots,$$

and this is less than $e^{|\beta_m|}$; hence we have

$$c_r\beta_m{}^r\left\{\frac{\beta_m}{r+1}+\frac{\beta_m{}^2}{(r+1)(r+2)}+\ldots\right\}=\theta_r\,|c_r\beta_m{}^r|\,e^{|\beta_m|},$$

where θ_r is some number whose modulus is between 0 and 1.

The modulus of

$$\sum_{r=p-1}^{r=np+p-1}\theta_r\,|c_r\beta_m{}^r|\,e^{|\beta_m|}\text{ is less than }e^{|\beta_m|}\sum_{r=p-1}^{r=np+p-1}|c_r\beta_m{}^r|,$$

or than

$$e^{|\beta_m|}\frac{|\beta_m|^{p-1}}{(p-1)!}C^{np+p-1}\{(|\beta_m|+|\beta_1|)(|\beta_m|+|\beta_2|)\dots(|\beta_m|+|\beta_n|)\}^p,$$

or than

$$e^{\bar\beta}\frac{\bar\beta^{p-1}}{(p-1)!}C^{np+p-1}\{(\bar\beta+|\beta_1|)(\bar\beta+|\beta_2|)\dots(\bar\beta+|\beta_n|)\}^p;$$

where $\bar\beta$ denotes the greatest of the numbers $|\beta_1|, |\beta_2|, \dots |\beta_n|$.

It thus appears that the modulus of

$$\sum_{r=p-1}^{r=np+p-1} c_r\beta_m^r\left\{\frac{\beta_m}{r+1}+\frac{\beta_m^2}{(r+1)(r+2)}+\dots\right\}$$

is less than a number of the form $PQ^p/(p-1)!$, where P and Q are independent of p and of m.

We have now

$$K_p(A+\sum_{m=1}^{m=n}e^{\beta_m})=K_pA+\sum_{m=1}^{m=n}\{\phi^{(p)}(\beta_m)+\dots+\phi^{(np+p-1)}(\beta_m)\}+L,$$

where K_pA is not a multiple of p, the second term is an integer divisible by p, and L is less than $nPQ^p/(p-1)!$. The prime p may be chosen so large that $nPQ^p/(p-1)!$ is numerically less than unity. Since $K_p(A+\sum_{m=1}^{m=n}e^{\beta_m})$ is expressed as the sum of an integer which does not vanish and of a number numerically less than unity, it is impossible that it can vanish. Having now shewn that no such equation as

$$A+e^{\beta_1}+e^{\beta_2}+\dots+e^{\beta_m}=0$$

can subsist, we see that πi cannot be a root of an algebraic equation with integral coefficients, and thus that π is transcendental.

It has thus been proved that π is a transcendental number, and hence, taking into account the theorem proved on page 50, the impossibility of "squaring the circle" has been effectively established.

CAMBRIDGE: PRINTED BY JOHN CLAY, M.A. AT THE UNIVERSITY PRESS

Made in the USA
Las Vegas, NV
10 February 2023

67277124R00039